# Key Concepts in Water Resource Management

The vocabulary and discourse of water resource management have expanded vastly in recent years to include an array of new concepts and terminology, such as water security, water productivity, virtual water, and water governance. While the new conceptual lenses may generate insights that improve responses to the world's water challenges, their practical use is often encumbered by ambiguity and confusion.

This book applies critical scrutiny to a prominent set of new but widely used terms, in order to clarify their meanings and improve the basis on which we identify and tackle the world's water challenges. More specifically, the book takes stock of what several of the more prominent new terms mean, reviews variation in interpretation, explores how they are measured, and discusses their respective added value. It makes many implicit differences between terms explicit and aids understanding and use of these terms by both students and professionals. At the same time, it does not ignore the legitimately contested nature of some concepts. The book will lead to greater precision on the interpretational options for the various terms, and for the value that they add to water policy and its implementation.

**Jonathan Lautze** is a researcher at the International Water Management Institute, based in its Pretoria office in South Africa.

# Earthscan Water Text series

**Key Concepts in Water Resource Management**
A Review and Critical Evaluation
*Edited by Jonathan Lautze*

**Contesting Hidden Waters**
Conflict Resolution for Groundwater and Aquifers
*By W. Todd Jarvis*

**Water Security**
Principles, Perspectives and Practices
*Edited by Bruce Lankford, Karen Bakker, Mark Zeitoun, Declan Conway*

**Water Ethics**
A Values Approach to Solving the Water Crisis
*By David Groenfeldt*

**The Right to Water**
Politics, Governance and Social Struggles
*Edited by Farhana Sultana, Alex Loftus*

# Key Concepts in Water Resource Management
## A Review and Critical Evaluation

Edited by
Jonathan Lautze

Routledge
Taylor & Francis Group

LONDON AND NEW YORK

from Routledge

First published 2014
by Routledge
2 Park Square, Milton Park, Abingdon, Oxon OX14 4RN

and by Routledge
711 Third Avenue, New York, NY 10017

*Routledge is an imprint of the Taylor & Francis Group, an informa business*

© 2014 International Water Management Institute

*British Library Cataloguing-in-Publication Data*
A catalogue record for this book is available from the British Library

*Library of Congress Cataloging-in-Publication Data*
Key concepts in water resource management: a review and critical evaluation/edited by Jonathan Lautze.
   pages cm.—(Earthscan water text)
   Includes bibliographical references and index.
   1. Water supply. 2. Water resources development. I. Lautze, Jonathan.
   HD1691.K526 2014
   333.91—dc23
   2013050048

ISBN: 978-0-415-71172-2 (hbk)
ISBN: 978-0-415-71173-9 (pbk)
ISBN: 978-1-315-88439-4 (ebk)

Typeset in Goudy
by Florence Production Limited, Stoodleigh, Devon, UK

# Contents

# Illustrations

## Figures

## Tables

# Contributors

**Xueliang Cai** is a water resources and irrigation engineer and remote sensing/GIS expert with ten years' international experience in Asia and Africa. He has been working on a range of issues related to water resources management for irrigated agriculture and ecosystems. Currently Cai plays leading roles in projects focused on i) small storages for integrated irrigation and aquaculture to enhance climate resilience in Malawi, Zambia, and Mozambique; ii) hydrological modeling of irrigation systems in Mozambique, Ethiopia, and South Africa to support community-led innovations; iii) assessment of hydrological functions of ecosystems in Zambezi and Volta basins. Previously Xueliang was also involved in a large effort "Global irrigated area mapping (GIAM)" which uses extensive remote sensing processing to extract irrigated areas at global, regional, and country level. He has conducted agricultural water productivity assessment for the Indus-Ganges, Limpopo, and Syr Darya basins. Xueliang obtained his Ph.D. degree from the Wuhan University, China, focused on water saving irrigation strategy development through system level integrated water resources management. Cai is engaged in academic and professional networks, and serves as an editor of the journal *Water International*.

**Sanjiv de Silva** is a Researcher at the International Water Management Institute based in his native Sri Lanka. Trained in Environmental Law, Sanjiv's abiding interest is in how governance mechanisms can support workable expressions of sustainable development that manage the interplay between varying resource–human systems. This has informed his 14 years as a practitioner and researcher designing and studying governance frameworks spanning a range of resource–human contexts across South and Southeast Asia. These include wildlife policy and law reform and implementation including participatory protected area management planning, integrated coastal resources governance and managing conservation-development trade-offs; water governance from different perspectives including government, small-holder, fisheries, hydropower, and environment, and the roles of gender, class, and caste in shaping climate change vulnerability and adaptation. Sanjiv has worked closely with policy makers and other government and non-state actors including local resource users. He was a Visiting Lecturer on Environmental Governance at the universities of Kelaniya and Sabaragamuwa, and is a member of the National Wetlands Steering Committee in Sri Lanka.

**Mark Giordano** is an Associate Professor of Environment and Energy in Georgetown University's School of Foreign Service. Mark is internationally known for his work on agricultural water management in developing countries, international water law, and conflict and cooperation over resources, and is recognized for his sometimes provocative views on natural resources management. He has more than 80 publications in major water science, geography, law, and international relations journals and advises a variety of organizations on water issues. Prior to joining Georgetown, Mark was a Managing Director of the International Water Management Institute (IWMI), one of the 15 Centers of the Consultative Group on International Agricultural Research. There he led the Institute's social science research program, focusing on improving agricultural water management for poverty reduction in Asia and Africa and influencing the global water agenda. Mark previously worked as a trade economist with the U.S. Department of Agriculture's Economic Research Service. He is from Eastern Washington but has spent his professional life in Austria, Botswana, Cambodia, China, Sri Lanka, Taiwan, and Zimbabwe.

**Munir A. Hanjra** is a development economist with over 20 years of professional experience on issues related to water resources such as global and regional water scarcity, food security, and poverty reduction. Currently a researcher at the IWMI—Southern Africa office, Dr. Hanjra has been involved in research and development programs on water, agriculture, and the environment in Australia, China, Canada, South and Southeast Asia, and Eastern and Southern Africa. He has published more than 40 scientific research papers in peer-reviewed journals and has made numerous other professional contributions. Current research interests include water and food security, water and poverty reduction, the water–food–energy nexus, wastewater reuse for nutrient capture and energy recovery, and climate change resilience and water sector adaptations.

**Jonathan Lautze** is a researcher at the IWMI—Southern Africa office in Pretoria. He has been involved in a range of applied research and development projects focused on topics such as water governance, water security, transboundary water management, climate change and water, and water and health. Lautze has published more than 20 peer-reviewed articles in journals and books, and has worked for the United States

Agency for International Development (USAID) and the World Bank. His development experience includes time living and working in Benin, Ethiopia, and Sri Lanka, and short-term assignments in Burkina Faso, Chad, Egypt, Ghana, Israel/Palestine, Jordan, Mali, Morocco, Oman, Rwanda, Senegal, Tajikistan, Turkey, Uganda, and Uzbekistan. He holds a Master of Arts in Law and Diplomacy from the Fletcher School, Tufts University and a Ph.D. from Tufts Department of Civil and Environmental Engineering.

**Herath Manthrithilake** serves as a Senior Researcher and Head of the research program of IWMI in Sri Lanka, titled "Sustainable Development Initiative." Prior to this assignment, Manthri lived and worked in India, and Central Asia, representing IWMI on its research projects. Manthri has experience in leading multidisciplinary and multicultural research teams in several countries. His Ph.D. (obtained in 1983) research was on stochastic hydrology and focused on "river flow regimes as Markov Chains; modeling and simulation for the management of river cascades— regulation, and use for agricultural and hydro power production." Subsequently, Manthri published a number of scientific papers on a variety of topics and has made presentations/communications on different aspects of water management in national, regional and international conferences. He possesses extensive experience serving on several National Science Committees in Sri Lanka—currently he serves on the Natural Resources Committee and in the Socio-economic Research Committee of the Council for Agricultural Research Policy (CARP), Water Resource Research Committee of National Science Foundation, Research and Training Committee of HARTI, Sri Lanka, Presidential Committee on Water Quality Management; among others.

**Greenwell Matchaya** is an Economist, currently an Economics Researcher at the IWMI, and Coordinator for the Regional Strategic Analysis and Knowledge Support System for Southern Africa, which is a multi CGIAR center program led by the International Food Policy Research Institute (IFPRI); under this program he coordinates efforts of various stakeholders toward strengthening analytical capacities of Southern African agricultural sectors, as one way of enhancing evidence-based policy planning and implementation in the Southern African Development Community. Dr. Matchaya has previously worked at the University of Reading (UK) where his research focused on examining the impact of agricultural policy on innovation/development, and also worked at the University of Leeds (UK) where his research focused on effects of health policy reforms in the National Health Service. Currently, his areas of interest are agricultural expenditure and water resources in development.

**Sanmugam A. Prathapar** has an international reputation in Agricultural Water Management and research management. He has led scientific, academic, and development institutions of international stature, demonstrating leadership skills with proven track record of success in formulating and implementing academic and research programs with vision and direction. During his 25 years of post-doctoral experience he has been a Researcher, Academic, Consultant, and a Development professional. He is now the Manager of the Groundwater Modelling group at the New South Wales Office of Water, Sydney, Australia. He has produced over 200 technical publications and presentations during his career, which include over 50 peer reviewed publications. Prior to joining IWMI, Dr. Prathapar was the Dean, College of Agricultural and Marine Sciences, Sultan Qaboos University, Oman; Professor at National Center for Groundwater Management, University of Technology, Sydney, Australia; and a Principal Research Scientist, CSIRO Division of Water Resources, Australia. He has held positions in Sri Lanka, USA, Australia, Pakistan, Oman, and India, and possesses research and development experience in South Asia and Sub-Saharan Africa.

**Luke Sanford** is a graduate student at the University of California–San Diego studying international economics, political science, and data analysis. He previously worked as an intern at the IWMI headquarters in Colombo, Sri Lanka in 2009–2010.

**Vladimir Smakhtin** has 30 years of experience as a Researcher and Research Manager in the broad area of water resources with an emphasis on global and regional water scarcity and food security. He holds a Ph.D. in Hydrology and Water Resources from the Russian Academy of Sciences. His research experience spreads across agricultural water management, environmental water management, low-flow and drought analyses and management, assessment of river basin development and climate change impacts on water availability and access, provision of hydrological information for data-poor regions, drought management, global water availability and scarcity. Vladimir was involved in a number of short and long-term assignments in Russia, Ukraine, Lithuania, South Africa, Canada, Sri Lanka, India, Pakistan, Afghanistan, Bangladesh, Vietnam, Cambodia, Iran, Morocco, Ethiopia, Ghana, Uzbekistan, Thailand, Laos. He consulted for a number of national and international government and non-government organizations including Department of Water Affairs of South Africa, Ontario Ministry of Natural Resources, World Commission on Dams, International Union for Conservation of Nature (IUCN), and United Nations Environment Proramme (UNEP). Dr. Smakhtin currently leads IWMI research programs on Water Availability and Access with a special focus on reducing water scarcity and ensuring livelihoods and food security globally.

**Aditya Sood** is a researcher at IWMI working on a wide range of topics dealing with watershed scale rainfall-runoff models, eco-hydrology, and integrated hydrological modeling. His areas of expertise include watershed and water quality management/modeling, GIS-based decision-support and planning, and integrated (water/energy/environment) policy analysis. Dr. Sood's research objectives include linking fundamentals of science, engineering, and policy in an interdisciplinary approach to understand hydrological processes as isolated processes and within complex eco-systems. He is interested in analyzing the impact of anthropogenic activities (such as land use changes and water usage) and of natural changes (such as climate change) on hydrological processes and vice versa. He has worked on various water related projects in Asia and Africa. He received his undergraduate degree in Civil Engineering, followed by a Masters in Environmental Engineering and Doctorate in Energy and Environmental Policy.

**Dennis Wichelns** is an agricultural and natural resource economist with many years of experience in production agriculture, and also in teaching, research, and policy analysis. He has served on the faculty of several colleges and universities in the United States, and as a Senior Fellow with the IWMI. While with IWMI, Dennis worked with many international colleagues on a variety of studies regarding water, agriculture, and livelihoods in several countries across Asia and Africa. He is currently one of several Editors-in-Chief of *Agricultural Water Management* and he is the founding Editor-in-Chief of *Water Resources & Rural Development*.

# Foreword

Don't we all need a *water secure* society where *water stresses* are properly dealt with through *sound governance structures*? And if IWRM is hard to implement, can we not then resort to a *Water-Energy-Food Nexus* or *virtual water*? If we are of a certain conviction, soft and *natural infrastructure* solutions can be by definition better than hard, engineering solutions. *Land grabs* are bad, *water grabs* even worse. Better use of *green water* is essential to reducing our *water footprints*.

Packaged concepts can come in handy, and using the same term to convey different ideas is not necessarily a bad thing if the concept is still evolving and an actual conversation is taking place. Nonetheless, there are an unfortunate number of concepts for which discussion has faded away, while confusion on substance has not.

Our "water box" is large enough to accommodate diverse arrays of thought, backgrounds, interests and preferences. Each combination of these results in a different perception of terms. When we step outside our water box and expose our terminology to those in sectors across which water cuts, we face deeper issues. Water terms may be associated with concepts in their respective "boxes". Water governance, for example, a multi-dimensional debatable notion for us water professionals, can be quite a challenge to explain to others. How about land and water governance? Environmental governance and water governance? Global governance for food, energy and water?

What can be additionally complicating—and lead to interesting preludes before real dialogues—is that explanations of fuzzy new terms may themselves contain unclear terms. In other words, terms that define terms may have multiple meanings. Food sector stakeholders may have substantially different perceptions of market and pricing mechanisms from water stakeholders, and safety nets can mean different things in the context of water services than in the context of energy services. And lack of shared understanding impedes our ability to constructively address issues in an optimal fashion.

*Key Concepts in Water Resource Management: A Review and Critical Evaluation* takes a critical and comprehensive approach to evaluating six prominent concepts (or sets of concepts) in water resources management, i.e. *water scarcity, water governance, water security, water productivity, virtual water and water footprints*, and *green and blue water*. We learn what they mean, how they originated and evolved, how they are interpreted and if and how we can measure them. Various definitions that exist are presented, with the source clearly spelled out: evolution of the term; concept or paradigm; and key differences between the definitions and how they play out in water management. Points of confusion and skepticism that exist are reported and explained; and uses, limitations and metrics are provided, with examples where applicable.

The book challenges the reader to think critically and rationally—based on evidence—to assess the meaning, role and utility of a concept. The terms covered in the book do not lend them to trivial, default definitions—even if many may think so. Despite frequent portrayal as *silver bullets* for water management, the terms of focus are not free from limitations, underlying assumptions, and sometimes flaws. These realities reinforce the need to gain a better handle on these terms to minimize potential for misunderstandings, misinterpretations, or miscommunication.

The book goes far beyond an annotated compendium of recent, prominent water concepts to clinically dissect six concepts (or sets of concepts) that have become central to the discourse of twenty-first century water management. Lautze and colleagues put such concepts through a fine filter to improve our understanding of them, and to generate broader insights about the process of new term introduction. The book fills a critical gap and will serve as a trusted reference to deciphering meanings and interpretations of major water concepts. I might actually propose the book be nicknamed: *Everything You Always Wanted to Know About Key Water Concepts (But Were Afraid to Ask)*.

I will conclude by warning against underestimation of this book's value and utility. While the authors rightfully convey that the book's primary focus is on water management in agriculture, those interested in other uses and aspects of water management will find the book equally useful. Likewise, those active in the "allied" sectors of water will find the book helpful in understanding a sector and its community that would like to be involved in the major decisions of the readers' respective sectors. We in the water sector often contend that the most important decisions related to water are taken outside the water sector, and that those involved in managing water should therefore inform those who make these decisions. If I were a decision maker in energy, food or the environment, I would certainly appreciate having this book around.

Olcay Unver
Deputy Director, Land and Water Division, FAO
Coordinator, UN World Water Assessment
Programme (2007–2013)

# Preface

The vocabulary and discourse of water resource management have expanded vastly in recent years to include an array of new concepts and terminology, such as water security, water productivity, virtual water, and water governance. While the new conceptual lenses may generate insights that improve responses to the world's water challenges, their practical use is often encumbered by ambiguity and confusion. *Key Concepts in Water Resource Management: A Review and Critical Evaluation* is the first attempt to systematically examine the value added of a set of new terms in water resources management.

This book applies critical scrutiny to a prominent set of new but widely used terms, in order to clarify their meanings and improve the basis on which we identify and tackle the world's water challenges. More specifically, the book takes stock of what several of the more prominent new terms mean, reviews variation in interpretation, explores how they are measured, and discusses their respective added value. It makes many implicit differences between terms explicit and will aid understanding and use of these terms by both students and professionals. It is hoped that the critical scrutiny contained in this book will lead to greater precision on the interpretational options for the various terms, and for the value that they add to the field of water management.

On a personal level, motivation for my involvement for this book was generated by participation in two projects, one on water governance funded by the United States Agency for International Development (USAID) and another on water security, funded by the Asian Development ment Bank. I was surprised to find how much time was spent simply trying to understand the central project terms "water governance" and "water security." Subsequent thought and discussion led to the realization that the ambiguity surrounding these two terms is far from unique, and that there are indeed frequent informal rumblings and sarcastic comments in hallways and corridors about the precise meaning and value added of a number of new terms introduced in water management such as water governance, green v. blue water, hydropolitics, etc. This spurred thoughts on why we are creating new terms and devoting substantial time and energy to deciphering their meanings, instead of tackling water management challenges with the terms and concepts that we already have at our disposal and that have proven their ability to address water management challenges.

To push toward clarity, the idea occurred to me to assemble several of the more prominent new concepts in the water management community in one text, and apply some critical analysis to each of them. When discussing this idea with colleagues, most were quite supportive, yet some questioned whether this idea might conform more closely to a dictionary than a full-fledged book along the lines I was describing—and along the lines you are now reading. After some deliberation, I came to the conclusion that it is relatively easy to achieve agreement on, or at least no objections to, fuzzy definitions to many of the new terms in the international water community, as suited for a dictionary. Challenges and differences of interpretation rapidly arise, however, when the meanings are unpacked to identify central components and applied to measure something.

It is therefore my hope that this book is much more than a dictionary of new water management terms. It does review definitions, but goes far beyond this by addressing divergence in interpretations, differences in methods of calculation when relevant, and highlighting consistencies and inconsistencies in usage. It is my hope that this will make many implicit differences between terms explicit, and foster progress by improving their use. It is also my hope that this will lead to greater clarity and precision on the interpretational options for the various terms, and for the value that they add.

This book is focused mainly on water management related to agriculture. The main chapters of the book examine a set of terms that have risen to prominence in agricultural water management. The book's target audience is people active in water management discussions, who make use of terms examined in this book. It is hoped that improved use of such terms by those actively involved in water management dialogues, will lead to improved use and interpretation of such terms more widely (e.g., by general public).

# Acknowledgments

I would like to acknowledge several colleagues who have given their time to review chapters of this book and express my sincere appreciation for their efforts. Such colleagues, some of whom are also authors on chapters that they did not review, include Upali Amerasinghe, Xueliang Cai, Mark Giordano, Bruce Lankford, Doug Merrey, Bharat Sharma, Vladimir Smakhtin, Charles Thompson, Richard Vogel, Kai Wegerich, and Dennis Wichelns. In addition, I would like to extend my sincere thanks to the authors who authored or co-authored chapters of the book. Finally, I would like to thank Charlotte Hiorns (Senior Project Manager at Florence Production) and Ashley Wright (Routledge) for their editorial expertise during the production of this book and also Sumith Fernando (IWMI Layout and Graphic Specialist) who designed and finalized the cover.

Jonathan Lautze

# Abbreviations

| | |
|---|---|
| ADB | Asian Development Bank |
| AfDB | African Development Bank |
| AWBA | Arizona Water Banking Authority |
| CAADP | Comprehensive African Agricultural Development Program |
| CARP | Council for Agricultural Research Policy |
| CGIAR | Consultative Group on International Agricultural Research |
| ET | evapotranspiration |
| FAO | Food and Agriculture Organization |
| GIAM | Global irrigated area mapping |
| GWP | Global Water Partnership |
| IDRC | International Development Research Centre |
| IFAD | International Fund for Agricultural Development |
| IFPRI | International Food Policy Research Institute |
| IUCN | International Union for Conservation of Nature |
| IWMI | International Water Management Institute |
| IWRM | Integrated Water Resource Management |
| MDG | Millennium Development Goal |
| MUS | Multiple Use Systems |
| NGO | Non-Government Organization |
| SEEAW | System of Environmental Economic Accounting for Water |
| SIDA | Swedish International Development Agency |
| TFP | Total Factor Productivity |
| UNDESA | UN Department of Economic and Social Affairs |
| UNDP | United Nations Development Programme |
| UNEP | United Nations Environment Programme |
| UNESCAP | United Nations Economic and Social Commission for Asia and the Pacific |
| UNESCO | United Nations Educational, Scientific and Cultural Organization |
| UNU | United Nations University |
| USAID | United States Agency for International Development |
| US EPA | US Environmental Protection Agency |
| USGS | United States Geological Survey |

| WA | Water Accounting |
| WA+ | Water Accounting Plus |
| WoK | Web of Knowledge |
| WP | Water Productivity |
| WWAP | World Water Assessment Program |
| WWC | World Water Council |

# 1   Introduction

*Jonathan Lautze*

## 1.1  Background

The water world has become saturated with new concepts such as water governance, water security, Integrated Water Resources Management (IWRM)—concepts that coexist with pre-existing notions such as water policy and institutions, and water management. While the new conceptual lenses may generate insights that improve responses to the world's water challenges, their practical use is often encumbered by ambiguity, confusion, and even fatigue associated with the steady flow of new "solutions" that can be interpreted in multiple ways. One result is lost time and energy devoted to sorting through the meanings and measuring sticks for a slew of new words, often in an ad hoc quick-and-dirty fashion, in order to achieve progress in deadline-oriented development projects. Another result of this ambiguity is the wild use of these terms in the context of policy discussions and development dialogues, which can distract people into sorting through meanings rather than sorting through issues and solving problems.

New vocabulary is nonetheless a fact of life. The flow of new terms is now a part of water management, will presumably continue to be, and may actually accelerate. It therefore makes sense not only to adapt to this trend, but to consider institutionalizing a process of adaptation to the steady stream of new terms. An initial step toward this end is to take stock of what several of the more prominent terms mean, to review variation in interpretation, to explore how they are measured, and to discuss their respective added value. Indeed, the current absence of a text that addresses these issues presents a rather major obstacle as many people—inside and outside the water community—have heard these terms yet lack a solid grasp of what they mean.

This book applies scrutiny to a prominent set of new terms in water management, in the hope that systematic application of critical thought can improve the basis on which we identify and tackle the world's water challenges. This book explores definitions of six prominent topics in water management, reviews their central components, and identifies tools used to measure them when applicable. Structurally, the book devotes one chapter to each of six new concepts that have entered—or grown greatly in prominence—in the water management community in the late twentieth or early twenty-first century. Chapters 2 through 5 each focus on understanding one particular term, and Chapters 6 and 7 package a few related terms together. A final chapter then synthesizes, and draws some lessons and recommendations.

Taken together, these chapters are intended to provide a guide to deciphering some of the prominent new concepts that permeate the water community. Our hope is that discussion of this set of terms together in a single text will allow it to serve as a general reference, and help to foster broader thought about the process of introducing new terms. The book purposely challenges certain notions that are often unquestioningly accepted, often in a provocative manner, in an attempt to remind us to keep thinking and questioning. It is hoped that others will subsequently apply some of the same critical approaches to the arguments contained in this book, in the process improving the precision and clarity surrounding many of the concepts that are discussed.

A tough choice when crafting this book was determining which terms to include for analysis. As noted above, there is no shortage of new terms in the water management community. Selection of the terms that are analyzed in this book was primarily determined by their level of prominence in the language of development agencies and international conference and fora, and by the level of confusion surrounding such terms. While IWRM was not explicitly included due to the volume of discussion already in existence on the topic (e.g., Molle, 2008), discussion of IWRM receives substantial attention in the chapter on water governance. The topics that were included are as follows:

- water scarcity
- water governance
- water security
- water productivity
- water footprints and virtual water
- green and blue water.

To be clear, the point of this book is not to invalidate the use of any of these terms. Rather, the point is to achieve greater clarity on what we mean by each of them, in order to create an improved basis for approaching real issues and challenges in the water management community. New terms and conceptual frameworks should be a means to gaining a better understanding of issues. They should not spur additional confusion that diminishes our understanding of issues. The expectation is that this book will be useful to water professionals seeking to better understand the new terms, researchers seeking to understand variations among the different terms and approaches to measuring them, and students wishing to gain an introduction to these often-used terms.

## 1.2 Overview of Chapters

The remainder of the book is divided into seven chapters and an appendix. Most chapters follow a fairly consistent structure in which definitions of the particular terms are reviewed, measures and indicators are compared, and value added is determined. Chapters were ordered roughly according to the level of prominence of their central term. Chapters 2 through 5—which include the concepts of water scarcity, water governance, water security, and water productivity—encompass four of the most common terms in twenty-first-century water management. IWRM, which may be the only term to challenge those just listed in frequency of use, was not the focus of a separate chapter since the concept receives coverage in the water governance chapter. Terms in Chapters 6 and 7—focused on virtual water and water footprints, and green and blue water—are growing in use, yet they have not achieved the same level of prominence as terms found in earlier chapters.

Chapter 2, co-authored by Jonathan Lautze and Munir Hanjra, is focused on water scarcity. Widely considered to present a major global challenge, water scarcity is an extremely prominent term in the water management community. Definitions of the term nonetheless vary and there are a range of indicators that measure scarcity in different ways. This chapter considers definitions and measures of water scarcity in the broader context of resources scarcity to identify consistencies and inconsistencies of usage. In particular, the chapter determines the degree to which prominent indicators of water scarcity—namely, the Falkenmark indicator, physical and economic water scarcity—align with notions of resources scarcity as a means to assessing the value added of the "true" definition and measure of water scarcity. The results reveal a conflation of the distinct concepts of water stress and water scarcity; i.e., that the former concept has become increasingly subsumed under the name of the latter. Isolation of water scarcity definition and indicators nonetheless reveals somewhat limited value. While the concept has helped raise the profile of water in development discussions, the practical use of the concept may benefit from sectoral disaggregation.

Chapter 3, co-authored by Jonathan Lautze, Sanjiv de Silva, Mark Giordano, and Luke Sanford, is focused on water governance. Water governance has emerged as perhaps the most important topic of the international water community in the early twenty-first century, and achieving "good" water governance is now a focus of both policy discourse and innumerable development projects. Somewhat surprisingly in light of this attention, there is widespread confusion about the meaning of the term "water governance." This chapter reviews the history of the term's use and misuse to reveal how the concept is frequently inflated to include issues that go well beyond governance. Further, it highlights how calls to improve water governance frequently espouse pre-determined goals—often derived from the tenets of IWRM—that should instead be the very function of water governance to define. To help overcome this confusion, the chapter considers the relationship between IWRM and water governance and suggests a more refined definition of water governance and related qualities of good water governance that are consistent with broader notions of the concepts.

Chapter 4, co-authored by Jonathan Lautze and Herath Manthrithilake, is focused on water security. "Water security" has come to infiltrate prominent discourse in the international water and development community, and achieving it is often viewed as a new water sector target. Despite the elevated status that the concept has increasingly acquired, understandings of the term are murky and quantification is rare. To promote a more tangible understanding of the concept, this chapter develops an index for evaluating water security at a country level. The index is comprised of indicators in five components considered to be critical to the concept: (i) basic needs, (ii) agricultural production, (iii) the environment, (iv) risk management, and (v) independence. Achieving water security in these components can be considered necessary but insufficient criteria to measure the achievement of security in related areas such as health, livelihoods, and industry. After populating indicators with data from Asia-Pacific countries, results are interpreted and the viability of methods is discussed. This effort comprises an important first step for quantifying and assessing water security across countries, which enables more concrete understanding of the term and discussion of its added value.

Chapter 5, co-authored by Jonathan Lautze, Xueliang Cai, and Greenwell Matchaya, is focused on water productivity. Improving water productivity (WP), especially in agriculture, is increasingly recognized as a central challenge in international development. A growing body of literature has nonetheless delimited the value and role of WP. This chapter compares WP with related concepts of water efficiency and agriculture productivity in order to interpolate particular benefits obtained through utilization of a WP perspective. The chapter's main finding is that WP holds value as a decision-making guide for allocating water between areas and sectors when applied with other indicators. The chapter also found that WP does not add value when applied in isolation in a particular location such as a scheme or farm; pre-existing indicators of water efficiency and agricultural productivity may in fact prove more useful at this scale. The chapter concludes by suggesting that "improving WP" should not be treated as a central challenge in water management, but the WP indicator holds value when employed together with other indicators.

Chapter 6, authored by Dennis Wichelns, is focused on virtual water and water footprints. The notions of virtual water and water footprints have gained considerable traction in scholarly literature and the popular press. The preponderance of articles on these topics might lead one to think the notions are based on a firm conceptual foundation and that they enhance understanding of challenging issues regarding water resources. This chapter reviews how these concepts are defined and measured, and identifies certain underlying flaws that greatly constrain the utility of those notions. Particular focus is devoted to dangers associated with direct application of virtual water to guide international trade. Further, the chapter reveals that comparing the water footprints of goods and services is an exercise that can easily mislead consumers into making decisions that inflict unintended harm on households and communities in faraway places. The chapter concludes that, while these two concepts help to shed important light on the role of water in trade, it is increasingly clear that use of such concepts in isolation could hold dangerous implications.

Chapter 7, co-authored by Aditya Sood, Sanmugam Prathapar, and Vladimir Smakhtin, is focused on green and blue water. Accounting for green water has received growing attention for its importance in reducing hunger, alleviating poverty and adapting to climate change. In particular, recognition for distinctions between green and blue water are presumed to unlock opportunities for improving water management in rainfed agriculture. Despite this attention, there is scant articulation of the value that the new paradigm has added relative to previously utilized concepts characterizing agriculture water use in the hydrologic cycle. Indeed, while the green v. blue water distinction may help reveal options for improving food security, it may be that other concepts could be equally used to achieve the same end. To understand the degree to which the green v. blue water paradigm has added value, this chapter compares this paradigm with classical approaches for conceptualizing water use in agriculture. Drivers and definitions for other water colors are also considered. The results of this analysis reveal that, while the reduction of water into simple colors may help to market certain concepts that might otherwise be perceived as esoteric, coloring water can also prove dangerously misleading.

Chapter 8, co-authored by Jonathan Lautze and Vladimir Smakhtin, is comprised of the book's conclusion. This chapter recapitulates findings as a means to generating guidance on how to move forward in a constructive fashion. The chapter first reviews underlying drivers for new term introduction, the value that they have added and sources of confusion associated with new terms. The chapter next derives broader lessons and recommendations on the process of new term introduction. Finally, the chapter offers thoughts on which concepts might serve as central paradigms or frameworks in water resources management if a thoughtful process were to be applied.

The final chapter of the book, co-authored by Munir Hanjra and Jonathan Lautze, is an Appendix that focuses on providing definitions and descriptions of 25 new terms in water management that are not contained in the book's main chapters. Approximately one paragraph is devoted to each term. Terms range from downstreamness to natural infrastructure to water accounting. The relative number of terms reviewed in the Appendix rendered this chapter more descriptive in nature.

## Reference

Molle, F. 2008. Nirvana concepts, narratives and policy models: Insight from the water sector. *Water Alternatives* 1(1): 131–156.

# 2 Water Scarcity

*Jonathan Lautze and Munir A. Hanjra*

## 2.1 Introduction

Water scarcity is widely considered to present a major global challenge (e.g., Seckler et al., 1998; Postel, 1998; WWC, 2001; UNDESA, 2007; Chartres and Varma, 2011; Vidal, 2012), often spurring language of a global "water crisis" (e.g., BBC, 2002; National Geographic, 2003; UN, 2006). Postel (1998), for example, questioned whether there will be enough water for food production in 2025. The World Water Council (WWC, 2001) described "the gloomy arithmetic of water." The UN Department of Economic and Social Affairs (2007) declared that water scarcity "will be among the main problems to be faced by many societies and the world in the 21st century." Chartres and Varma (2011) state that the world faces an emerging water crisis due to worsening water shortage and scarcity.

While mention of water scarcity is ubiquitous in the international development and water management communities as well as the popular press, inconsistencies, anomalies, and limitations associated with the definition and use of the water scarcity concept are apparent. Brown and Matlock (2011) document the considerable variation in how water scarcity is defined, interpreted, and measured. Rijsberman (2006) reveals inconsistencies associated with various water scarcity indicators, and questions whether water scarcity is fact or fiction. Molle (2008) suggests that demand for water will always outpace supply, hence scarcity is fairly widespread and the practical value of the concept is minimal. Rogers (2008) and Rogers and Leal (2010) question whether we are really facing a scarcity-induced crisis. The UN Human Development Report (2006) suggests that water scarcity is "manufactured through political processes and institutions that disadvantage the poor," a description consistent with Mehta's (2007) declaration that water scarcity is a socially mediated construct. Finally, Savenije (2000) highlights how inclusion of green water can skew water scarcity results, and Brandt and Vogel (2008) and Perveen and James (2011) highlight how variation in scale of analysis can affect results.

These documents highlight limitations of the water scarcity concept and variation in water scarcity indicators. However, they stop short of comparing variation in the interpretation of water scarcity with broader notions of resources scarcity. To provide an improved framework for assessment and discussion of water scarcity, this chapter considers definitions and measures of water scarcity in the broader context of resources scarcity as a means to understanding which interpretation of water scarcity is most consistent with resources scarcity and hence most fundamentally sound. The chapter first reviews broader notions of natural resources scarcity (section 2.2). The chapter then identifies definitions of water scarcity (section 2.3), and examines how these definitions have led to the creation of various indices —namely, the Falkenmark indicator, water stress, and economic water scarcity—with conflicting implications and interpretations (section 2.4). The degree to which various definitions and indices align with broader notions of resources scarcity is then examined (section 2.5), and the value added of approaches determined to be most conceptually sound are considered (section 2.6). A conclusion (section 2.7) then explores the value and role of water scarcity.

## 2.2  Background: Natural Resources Scarcity

Scarcity is defined as the fundamental problem of having humans who have unlimited wants and needs in a world of limited resources (Gregory, 2004). From a conventional economic perspective, scarcity occurs when demand exceeds supply at a given price (Hall and Hall, 1984). For goods bought and sold in a conventional market, the response to scarcity is for prices to rise such that demand and supply are at equilibrium. A good traded at a higher relative price, therefore, is scarcer than a good selling at a lower relative price. Notably, four assumptions underlying this perfect competition approach are that i) the good in question is non-essential and homogenous, ii) the good has close substitutes, iii) there are many firms and they are able to easily enter or exit the market, and iv) consumers have complete information on prices, quantities etc. If the price of one brand of shirt grows too high, for example, a consumer can exit the market or purchase another brand instead, as the good in question is non-essential, has close substitutes, many sellers, and available at competitive prices.

While scarcity of natural resources possesses many elements in common with conventional economic notions of scarcity, there are at least three important differences. First, many natural resources are not bought, sold, or traded in free markets and hence their allocation is not directly determined by market forces. This existence of imperfect market, pseudo prices and social–political factors determining allocations for many natural resources creates conditions in which some degree of "scarcity" is virtually always present. Second, natural resources generally possess attributes of public goods such that free markets are normally unable to determine their true prices and ensure efficient resource allocation. Third, there are perceptions of binding physical constraints on the availability of many natural resources. Unlike many conventional products on the market, for example, it is asserted that one cannot simply produce more land, air, or water to satisfy growing human demand.

Parameters specific to natural resources scarcity have engendered high-profile debates about how growing resources demand, driven by growing population, would spur global crises. Building on the logic of Malthus (1898), Ehrlich (1968) asserted that growing populations would lead to levels of demand for natural resources that far outstripped their availability, triggering a "population bomb." Simon (1980) disputed this assertion, suggesting that rising demand for natural resources would be outpaced by technological advances and productivity improvements; his argument was consistent with that of induced innovation (Boserup, 1965). Several others have weighed in on this debate (e.g., Barnett, 1979; Krautkraemer, 2005), adding nuance to the original arguments on both sides.

Importantly, at least two assumptions appear to underlie this debate. A first assumption is that demand for certain natural resources can be determined as a function of population and development levels rather than actual use. In a sense, broader notions of scarcity were tailored to natural resources by specifying innate levels of "wants and needs" necessary for survival. Second, demand for certain natural resources—particularly those that are rarely traded—may not always be reflected in the use of those resources. Indeed, a crisis was envisioned to result from the mismatch between demand for resources required to satisfy fundamental requirements and the quantity of those resources actually used, which would presumably be constrained by natural limits on supply.

Predictions of crises that would result from resources scarcity have generally not materialized. While the price of energy has gone up and certain regions face water shortages, often of a seasonal nature, food production has grown dramatically as a byproduct of productivity increases. The fact that crises have largely not resulted from natural resources scarcity is worth noting, and spurs questions related to the value that the debate over natural resources scarcity has contributed. While it may be that scarcity debates increased public awareness and prompted policy changes that averted crisis, it may also be that natural processes of induced innovation (e.g., productivity improvements) would have occurred regardless of such debates.

## 2.3 Defining Water Scarcity

Building on broader notions in natural resources scarcity and the related focus on sustainability (WCED, 1987), interest in the scarcity of water intensified in the 1990s and 2000s. Reviewing literature on water scarcity reveals two common definitions. The first definition of water scarcity is as follows (Rijsberman, 2006; IWMI-CA, 2007; Chartres and Varma, 2011):

When an individual does not have access to safe and affordable water to satisfy her or his needs for drinking, washing or their livelihoods we call that person water insecure. When a large number of people in an area are water insecure for a significant period of time, then we can call that area water scarce.

The second definition, which may be more consistent with broader notions of resources scarcity, is as follows (UN-Water, 2006; UN-Water and FAO, 2007):

[Water scarcity is] the point at which the aggregate impact of all users impinges on the supply or quality of water under prevailing institutional arrangements to the extent that the demand by all sectors, including the environment, cannot be satisfied fully.

The first definition marks an important way to reflect the water limitations faced in many parts of the world. It is straightforward and intuitive: if people lack access to water in a region, that region is water scarce. The definition nonetheless fails to distinguish whether lack of access to water results from: i) mismanagement, infrastructure limitations or lack of investments, or ii) limitations of water itself. The second definition presents a more nuanced, yet somewhat more esoteric, interpretation of the water scarcity concept. It considers scarcity to occur when demand for water exceeds the supply of water. While the definition is carefully worded and conceptually sound, it is nonetheless unclear about certain details necessary for application of this concept; namely, i) how water supply and water demand are to be determined, and ii) thresholds associated with degrees of scarcity, which are critical because water's public good nature and lack of price mean that water demand is likely to always be greater than supply.

In addition to the two definitions outlined above, a distinction between *physical water scarcity* and *economic water scarcity* has been elaborated by IWMI. Whereas *physical water scarcity* is considered to occur when "available water resources are insufficient to meet all demands," *economic water scarcity* is considered to occur "when investments needed to keep up with growing demand are constrained by financial, human, or institutional capacity" (Rijsberman, 2006; IWMI-CA, 2007). In essence, physical water scarcity would appear quite consistent with the UN-Water definition elaborated above, focused on the point at which demand exceeds or impinges on supply. Economic water scarcity, by contrast, looks beyond physical availability by incorporating the notion of water access. In other words, countries that have sufficient renewable water resources, but lack investment to tap those resources for human use, are defined as *economically water scarce* (Rijsberman, 2006; IWMI-CA, 2007).

The distinction between physical and economic scarcity draws attention to the fact that even in areas that are theoretically water abundant, water may not always be practically available to humans due to limitations on infrastructure, institutions, or other capacity. There is indeed an important distinction that should be recognized between water that is theoretically available for human use—e.g., renewable water resources—and water that is practically available for human use as a result of effective infrastructure, institutions, and capacity. By shedding light on this important difference, the physical v. economic distinction helps to expose an important nuance.

The conceptualization of a distinct economic water scarcity concept nonetheless appears questionable, for at least two reasons. First, the extent to which economic water scarcity reflects a scarcity of water is debatable. The concept is indeed predicated on the premise that there is sufficient (read: not scarce) water, but insufficient investment to tap and distribute that water. Second, it is possible for a region to be both physically and economically water scarce. That is, certain regions or countries may face physical water scarcity due to natural constraints on water availability (e.g., low renewable water resources), and such regions may also face economic water scarcity due to conditions in which only a small subset of available supplies are developed as a result of low investment. In such regions, it is not clear whether physical water scarcity supersedes economic water scarcity, or vice versa.

## 2.4 Measuring Water Scarcity to Produce Global Maps

Perhaps more important than the definitions outlined above are the water scarcity indicators (e.g., Falkenmark, 1989; Alcamo et al., 1997; Alcamo and Henrichs, 2002; Smakhtin et al., 2004; IWMI-CA, 2007; UNEP, 2008) that are widely applied to generate global maps that highlight a presumed water crisis. Water scarcity indicators can be broadly divided into three groups according to the ratio or fraction on which they are conceptually based: 1) water supply-to-water demand, 2) water use-to-water supply, 3) human-water-demand-to-human-water-access (Table 2.1). To understand the range in methods used to produce water scarcity maps, this section reviews the basic formula of a prominent indicator in each of the three groups and presents global maps that have resulted from their use.

### Scarcity as Water Supply-to-Demand: The Falkenmark Indicator

The most prominent indicator in the first group, focused on water supply to water demand, is the Falkenmark indicator (Falkenmark, 1989; Falkenmark et al., 1989). Other notable indicators in this group include Ohlsson (1999) and Sullivan et al. (2003), which incorporate a capacity to cope with scarcity. The Falkenmark indicator consists of only two variables: i) renewable water resources as a marker of water supply, and ii) human population as a proxy for water demand. Water supply is divided by

*Table 2.1* Methods of calculating water scarcity

| Water scarcity indicator group | Primary example | Method of calculation |
| --- | --- | --- |
| 1. | Falkenmark indicator | Renewable water resources/population |
| 2. | Physical water scarcity and water stress indicator | Withdrawal/renewable water resources |
| 3. | Economic water scarcity | High malnutrition (proxy for unsatisfied water demand) and (withdrawal/renewable water resources) < 25% |

population to determine water availability per capita per year: 1,700m³ of renewable water resources per capita per year is the threshold for satisfying water requirements in the household, agricultural, industrial and energy sectors, and the needs of the environment (Figure 2.1). Regions that cannot achieve such a ratio are said to experience *water stress*, and regions that fail to achieve 1,000 m³ per capita are said to experience *water scarcity*.

The logic of the Falkenmark indicator appears conceptually aligned with broader notions of resources scarcity. The indicator is based directly on the relationship between supply of and demand for water. Further, the indicator is relatively transparent and easy to interpret. The indicator's treatment of water demand as solely population-driven nonetheless warrants more scrutiny. While such a relationship may in fact hold if focus is devoted only to domestic water demand, determining a region's aggregate water demand as a direct function of population is misleading. Local demand for agriculture goods does not require local production of such goods and may not translate into associated local demand for water, for example, since it is possible to satisfy demand for agriculture goods through trade. Treating water demand as a direct correlation of population becomes additionally misleading when such a correlation is used to project future demand for water—as done by the creator of this indicator and others (e.g., Falkenmark, 1989). Indeed, temporal changes in water use efficiency, policy choices, institutional change, and technological advances may affect demand for water— particularly demand for agricultural water—in such a way that mitigates impacts of population change (e.g., Gleick, 2002).

### Scarcity as Water Use-to-Water Supply: The Water Stress Indicator

A prominent indicator in the second group of water scarcity indicators, focused on water use relative to water supply, is the water stress indicator

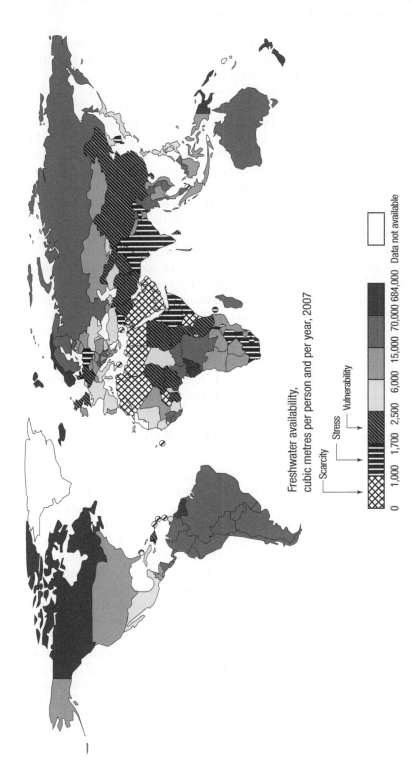

Freshwater availability,
cubic metres per person and per year, 2007

Scarcity
┌─ Stress
│  │  ┌─ Vulnerability
↓  ↓  ↓

0   1,000  1,700  2,500  6,000  15,000  70,000  684,000   Data not available

*Figure 2.1* Water scarcity based on the Falkenmark indicator.

Source: UNEP, 2008.

(Alcamo et al., 2000; Alcamo and Henrichs, 2002). Although this indicator was initially combined with additional indicators to form a broader index that measures vulnerability or scarcity (Raskin et al., 1997; Alcamo et al., 1997), it was later used alone. The indicator consists of two variables: i) water withdrawal for human use, and ii) total renewable water resources. Water withdrawal is divided by renewable water resources to produce a ratio that gauges water stress (Alcamo and Henrichs, 2002). IWMI (IWMI-CA, 2007) has since used the same approach to measure *physical water scarcity*. According to the water stress indicator (Figure 2.2), a region faces low water stress if annual withdrawals are below 20 percent of available water supply, mid stress if withdrawals are between 20 and 40 percent of annual supply, and high stress if this figure exceeds 40 percent.

While this indicator is intuitive and easy to interpret, the degree to which it measures water scarcity is again questionable. Indeed, treating the water stress indicator as a measure of water scarcity rests on two fallacious assumptions: i) water demand can be equated to water use, ii) water use can be reduced to water withdrawal. The first assumption is dubious because, as noted above, demand for water can exceed its use—it is precisely such a gap between demand and use that is presumed to trigger a crisis. The second assumption is misleading since water withdrawal comprises just one type of water use. Many regions rely heavily on production from water use in rainfed agriculture, for example, for which water is typically not withdrawn. Further, the degree to which water reuse is incorporated is not clear.

### Scarcity as Human-Water-Demand-to-Human-Water-Access: Economic Water Scarcity

The third group of water scarcity indicators, focused on human-water-demand-to-human-water-access, is only known to include the economic water scarcity indicator (IWMI-CA, 2007). The conceptual basis for this indicator appears to be that human demand for water either is or is not equal to human access to water. Regions where human demand and access are not equal—i.e., where human demand for water exceeds human access to water—are classified as economically water scarce. Determination that human water demand exceeds access is made by interpolating that water withdrawal is too low to meet human needs. Determination that water withdrawal is insufficient to satisfy human needs, in turn, is made based on satisfaction of two criteria: i) water withdrawal is low, quantified as withdrawal less than 25 percent of renewable water resources, and ii) demand for water is not completely met, revealed by high rates of malnutrition. The result of applying the economic scarcity indicator, as well as the physical scarcity indicator which is identical to the water stress indicator elaborated above, is shown in Figure 2.3.

The economic water scarcity indicator sheds light on the fact that renewable water resources are not always readily accessible for human use.

*Figure 2.2* Water stress as water withdrawal to water availability.
Source: Alcamo and Henrichs, 2002.

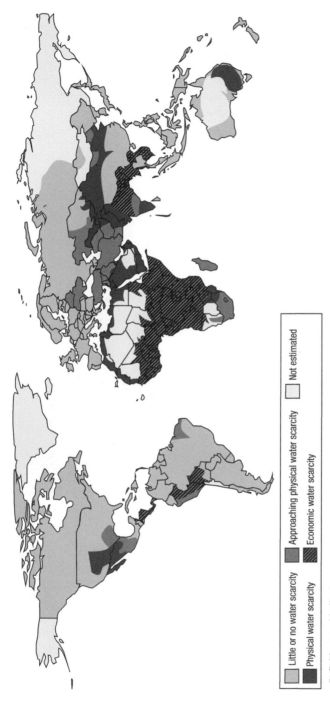

**Definitions and indicators**

- *Little or no water scarcity*. Abundant water resources relative to use, with less than 25% of water from rivers withdrawn for human purposes.
- *Physical water scarcity* (water resources development is approaching or has exceeded sustainable limits). More than 76% of river flows are withdrawn for agriculture, industry, and domestic purposes (accounting for recycling of return flows). This definition—relating water availability to water demand—implies that dry areas are not necessarily water scarce.
- *Approaching water scarcity*. More than 60% of river flows are withdrawn. These basins will experience physical water scarcity in the near future.
- *Economic water scarcity* (human, institutional, and financial capital limit access to water even though water in nature is available locally to meet human demands). Water resources are abundant relative to water use, with less than 25% of water from rivers withdrawn for human purposes, but malnutrition exists.

Legend:
- Little or no water scarcity
- Approaching physical water scarcity
- Not estimated
- Physical water scarcity
- Economic water scarcity

*Figure 2.3* The IWMI water scarcity map.

SOURCE: IWMI-CA, 2007.

That is, the indicator exposes the fact that a region may possess high water availability in theory, yet practical availability of that water for human use may be limited due to lack of investments. The indicator's assumption that high malnutrition results from low water withdrawal—and the implied corollary that increases in withdrawal may alleviate malnutrition—may nonetheless be misleading. This singular focus on withdrawal would indeed appear to belittle the role that improvements in income derived from non-water-intensive sectors (e.g., industry) may play in reducing malnutrition. Further, while withdrawal-driven increases in agricultural production may boost incomes, which in turn reduce malnutrition, withdrawal is but one among many means to raise incomes derived from agricultural production. Provision of improved seeds and fertilizer, for example, or facilitation of improved market access, may accomplish the same end with constant levels of withdrawal.

## 2.5   Results: Determining Alignment and Identifying Inconsistency

A comparison of definitions of natural resources scarcity with the two definitions and various measures of water scarcity reveals which water scarcity definition and indicator is most consistent with the commonly accepted definitions of resources scarcity. Through its explicit focus on the point at which demand for water exceeds supply of water, the UN-Water definition is most consistent with broader conceptualizations of resources scarcity. Similarly, through its explicit focus on water demand and water supply, the Falkenmark indicator appears most aligned with notions of resources scarcity. The Falkenmark indicator effectively translates fundamental notions of demand for a resource and supply of a resource into specific water terms, and explicitly prescribes how to quantify those terms at a country level. While it is fair to debate precisely how such terms are quantified in the context of this indicator, particularly in the case of water demand, the fact remains that the two fundamental variables contained in the notions of resources scarcity can be directly matched with the two measurable variables in the Falkenmark indicator.

In contrast to the Falkenmark indicator, the central variables in the water stress indicator do not enjoy the same degree of alignment with broader notions of natural resources and water scarcity. To reveal the mismatch, it is helpful to directly juxtapose the two main variables in the water stress indicator with the two main variables in broader definitions to enable comparison of the degree to which water withdrawal is consistent with water demand, and the degree to which renewable water resources are consistent with water supply. While ascribing consistency between renewable water resources and water supply would likely trigger few objections, the degree to which water withdrawal is consistent with water demand is far more

debatable. As noted, broader notions of natural resource scarcity are predicated on the basis that demand for water and natural resources is not always
manifested in use; indeed, the disparity between demand and use spurred
debate about the potential for a crisis. As such, it is conceptually dubious
to equate water demand with water withdrawal in the context of water
scarcity.

The anomaly of equating water demand with water use can be highlighted
with a pair of examples. A first example may be evidenced in certain poor
regions of the world, where populations utilize only 20 liters/person/day
for their domestic needs. Can we treat this quantity as their demand for
domestic water? Or do we equate domestic water demand with an internationally recognized quantity (e.g., 100 liters/person/day) and assume that
there is unsatisfied demand and therefore a distinction between use and
demand? The logical basis of the latter question actually coincides with that
of the economic water scarcity concept, discussed in more detail below.
A second example could come from Central Asia. One can state fairly
uncontroversially that environmental water demands are not met here, due
largely to excessive use of water for agriculture. Yet can we equate this
high—some would say unsustainable—use of agricultural water to water
demand? If so, we would introduce a good degree of subjectivity to the
determination of demand, and grandfather in behavior that prioritizes
demands by one sector at the expense of another.

To be clear, water stress is a valid concept, and the water stress indicator
can reveal useful information—potentially more useful than information
obtained through use of the Falkenmark indicator. Further, the indicators
of water stress likely embody a greater degree of sophistication than those
of the Falkenmark indicator. Nonetheless, water stress appears conceptually distinct from the concept of water scarcity. While the motivations
underlying the indicator's name-migration from stress to scarcity are not
entirely clear, what is clear is that the conflation and confusion resulting
from loose use of terms describing this indicator and others may undermine
their practical application in appropriate roles to tackle real challenges.

A final issue that merits focus is the degree of conceptual alignment
between the economic water scarcity indicator and broader notions of
scarcity. At its conceptual core, economic water scarcity is focused NOT
on water demand relative to water supply, as specified in the definition, but
rather on water demand relative to water access. Indeed, when water demand
is determined to be equal to water access, a region is presumed to not experience economic water scarcity; when water demand exceeds water access, a
region is determined to be economically water scarce. As such, while creators
of the economic scarcity may have been well-intentioned, this measure of
water scarcity is clearly at odds with broader notions of natural resources
and water scarcity.

## 2.6   Discussion: What value? What role?

This chapter has presented definitions of natural resources and water scarcity, as well as various measures of water scarcity. The chapter has determined that, of the two major definitions of water scarcity, the UN-Water interpretation is most consistent with broader conceptualizations of natural resources scarcity. Further, the chapter compared fundamental notions of scarcity elaborated in these definitions with the various indicators used to measure water scarcity, in order to identify which indicator is most aligned with broader notions of the concept and therefore most conceptually sound. The Falkenmark indicator was determined to be the most suitable according to this standard.

Determination of the most appropriate definition and measure of water scarcity in turn triggers a suite of broader follow-on questions, such as: What practical value has the concept added, and in what role is it suitable to be applied? The value and role of water scarcity, and the Falkenmark indicator used to measure it, may in fact mirror that of natural resources scarcity. It should be recalled that natural resources scarcity rose in prominence largely in the context of alerting us to a crisis that was presumed to be looming. This rise in conceptual prominence helped to foster the development of various indicators of natural resources scarcity (e.g., Hall and Hall, 1984; Krautkraemer, 2005). While the crisis has not materialized, it is difficult to determine whether i) scarcity discussions engendered policy shifts that averted such a crisis, or ii) such a crisis would have been otherwise averted due to natural processes of adaptation and innovation.

Similar to discussions surrounding natural resources scarcity, the value and practical use of the Falkenmark indicator often finds its way into roles of highlighting doomsday conditions and crises that are envisioned to come. Demand for water is projected forward to highlight that population increases will lead to a much greater demand for water, which is presumed to cause a future water crisis. Such crisis predictions often find their way into initial sections of articles and reports (e.g., Chartres and Varma, 2011), and are used in such a way to justify a more prominent focus on water management. Nonetheless, linkages between doomsday projections of scarcity and specific issues in articles and reports are often tenuous.

A broader issue with the water scarcity concept and indicator is that the implications of its message may be at odds with the purpose of employing the concept. That is, if a central goal in employing this concept is highlighting the urgency of a looming water crisis as a means to justifying the need for a specific water project, questions should be posed about the implications of determining that a region is water-scarce. Indeed, while responses to conditions of water scarcity may include improving our management of water through policy reforms and investments so that less is wasted, responses to scarcity conditions can also include implementation of many non-water measures—improving seed varieties, adding fertilizer, improving access to markets and post-harvest food storage—to increase the output achieved through water use so that more is produced with the same quantity of water consumption. In short, if a primary purpose of applying water scarcity indicators is to galvanize support for water sector projects, it is worth noting that many of the responses to water scarcity lie outside the water sector. Moreover, the Falkenmark indicator may in fact mask certain practical water-sector options—for example, increasing the use of un-used water—for improving water management in a region.

While conclusive determination of explanatory factors is not possible, one suspects that the increasing popularity of the water scarcity concept may have led distinct concepts and indicators to adopt scarcity language. As noted above, the water stress *water use-to-water supply* indicator gradually migrated from a measure of vulnerability and stress into a measure of water scarcity. This transition likely reflects a desire to subsume the valid yet distinct concept of water stress under the presumably trendier heading of water scarcity. A similar point could be made related to economic water scarcity. The concept essentially indicates that a region possesses insufficient infrastructure to tap water—water that is presumably available and not scarce. Yet, even in this case, there is a contrived attempt to couch non-scarcity in the language of scarcity. Indeed, if a region's infrastructure is too scarce to enable sufficient water use, why not simply call that region infrastructure-scarce?

The bottom line is that the practical contribution of the water scarcity concept and the Falkenmark indicator is unclear. Like broader projections of crises associated with natural resources scarcity, doomsday predictions of scarcity-induced water crises likely succeeded in raising awareness and engendering concern—which in turn may have helped lead to investments that alleviated certain crisis drivers. Nonetheless, recent declarations that an impending water crisis is more related to water governance than scarcity (e.g., UNESCO, 2006) may actually reflect recognition for the limitations of message of a scarcity-induced water crisis. Indeed, whether or not this shift away from scarcity-induced crisis to governance-induced crisis was a calculated one, it certainly works to direct more attention to in-house water sector policy options. If the goal of employing concepts is to attract attention and support to the water sector, it makes sense to highlight challenges—for example, poor water governance and management, high rainfall variability, insufficient storage—that require solutions internal to the water sector.

## 2.7   Conclusion: Disaggregate Sectors

This chapter has reviewed definitions and measures of natural resources scarcity and water scarcity. It identified the UN-Water definition of water scarcity to be most consistent with broader notions of resources scarcity, and determined the Falkenmark indicator to be most consistent with definitions of both natural resource and water scarcity. Discussion was then devoted to the value that the water scarcity concept, and application of the Falkenmark indicator, have added. This discussion revealed that the practical value of this concept, and application of the indicator, is unclear. The discussion also postulated that there is an allure associated with the term water scarcity that has likely spurred distinct concepts to incorporate the language of scarcity.

One issue that may not have been addressed in this document relates to consideration of water scarcity—and more particularly water demand—in different water-using sectors. The Falkenmark indicator appears to consider water demand as an innate human requirement, with demand for water in a region treated as a direct function of population in that region. While water demand in the domestic sector could be tied to population size, demands in other sectors is likely more closely linked to factors other than population. In other words, whereas water requirements for domestic use could be said to be a function of population (e.g., according to an international standard such as 100 liters/person/day, as set by the WHO), demand for water in other sectors such as agriculture is far less straightforward. A country's population-based demand for agricultural goods—and the water implicitly embedded in such goods—does not require in-country demand for such water. Indeed, goods have been traded for ages in order to exploit comparative advantages and satisfy local demand with imported supplies.

Limitations associated with applying the Falkenmark indicator to non-domestic sectors may call for greater attention to the rationale for application of the indicator to different sectors. In essence, the level of water use in a region is often a function of policy choices, and social and economic factors. To treat levels of water use that result from such choices and factors as water demand would be to accept that water demand—and hence the water scarcity that follows when demand exceeds supply—is a manufactured process, as suggested by some (e.g., UN, 2006; Mehta, 2007). Nonetheless, as established above, it is fairly dubious to equate water use with water demand. These realities engender fairly profound questions about the degree to which water scarcity is a concept that can validly be applied to non-domestic sectors of water use.

In conclusion, it may make sense to disaggregate sectors that have been traditionally aggregated in water scarcity indicators. Sectoral disaggregation would likely defuse many of the apocalyptic scenarios of water crisis associated with water scarcity, with alarmist language of "absolute scarcity" supplanted by more sober descriptions such as "plenty of water for drinking, cooking, cleaning, but most food will need to be imported." Notably, few regions would appear water scarce when assessed solely from the perspective of their domestic water requirements because domestic water requirements are quite low compared to those of other sectors. Nonetheless, the relative authority with which one can apply water sector indicators to the domestic sector—and the numerous confounding variables present when one applies water scarcity to a broader set of sectors—may call for confining application of the water scarcity indicator to the domestic sector.

## References

Alcamo, J., and Henrichs, T. 2002. Critical regions: A model-based estimation of world water resources sensitive to global changes. *Aquatic Sciences* 64: 1–11.

Alcamo, J., Doll, P., Kaspar, F., and Siebert, S. 1997. Global change and global scenarios of water use and availability: An application of WaterGAP 1.0. Kassel: University of Kassel, CESR.

Alcamo, J., Henrichs, T., and Rosch, T. 2000. World water in 2025: Global modeling and scenario analysis, in: F. Rijsberman (Ed.) *World Water Scenarios Analyses*, Marseille: World Water Council.

Barnett, H. 1979. Scarcity and growth revisited. In: V. K. Smith (Ed.) *Scarcity and Growth Reconsidered*. Baltimore, MD: Johns Hopkins University Press for Resources for the Future, 163–217.

BBC. 2002. UN warns of looming water crisis. Available online: http://news.bbc.co.uk/2/hi/1887451.stm

Boserup, E. 1965. *The Conditions of Agricultural Growth: The economics of agrarian change under population pressure*. London: Allen & Unwin.

Brandt, S., and Vogel, R. 2008. Indicators of hydrologic stress in Massachusetts. Hawaii: ASCE-EWRI, World Water & Environmental Resources Congress.

Brown, A., and Matlock, M. 2011. A review of water scarcity indices and methodologies. White Paper #106. The Sustainability Consortium. University of Arkansas.

Chartres, C., and Varma, S. 2011. *Out of Water: From abundance to scarcity and how to solve the world's water problems*. New York: FT Press.

Ehrlich, P. 1968. *The Population Bomb*. New York: Ballantine Books.

Falkenmark, M. 1989. The massive water scarcity facing Africa now—why isn't it being addressed. *Ambio* 18(2): 112–118.

Falkenmark, M., Lundquist, J., and Widstrand, C. 1989. Macro-scale water scarcity requires micro-scale approaches: Aspects of vulnerability in semi-arid development. *Natural Resources Forum* 13(4): 258–267.

Gleick, P. 2002. *The World's Water: The biennial report on freshwater resources 2002–2003*. Washington DC: Island Press.

Gregory, P. 2004. *Essentials of Economics*. Sixth Edition. Chicago, IL: Cengage Learning.

Hall, D., and Hall, J. 1984. Concepts and measures of natural resource scarcity with a summary of recent trends. *Journal of Environmental Economics and Management* 11: 363–379.

IWMI-CA. 2007. *Water for Food, Water for Life: A comprehensive assessment of water management in agriculture*. Colombo: International Water Management Institute, London: Earthscan.

Krautkraemer, J. 2005. Economics of natural resources scarcity: The state of the debate. Discussion Paper 5, 14. *Resources For the Future*. Washington, DC.

Malthus, T. 1898. *An Essay on the Principle of Population*. London: Penguin Books.

Mehta, L. 2007. Whose Scarcity? Whose property? The case of water in western India. *Land Use Policy* 24: 654–663.

Molle, F. 2008. Why enough is never enough: The societal determinants of river basin closure. *International Journal of Water Resources Development* 27(2): 217–226.

National Geographic 2003. UN highlights world water crisis. Authored by Hillary Mayell. Available online: http://news.nationalgeographic.com/news/2003/06/0605_030605_watercrisis.html

Ohlsson, L. 1999. Environment, scarcity and conflict: A study of Malthusian concerns. Göteborg: University of Göteborg, Department of Peace and Development Research.

Perveen, S., and James, L. 2011. Scale invariance of water stress and scarcity indicators: Facilitating cross-scale comparisons of water resources vulnerability. *Applied Geography* 31: 321–328.

Postel, S. 1998. Water for food production: Will there be enough in 2025? *BioScience* 48(8): 629–637.

Raskin, P., Gleick, P., Kirshen, P., Pontius, G., and Strzepek, K. 1997. *Water Futures: Assessment of long-range patterns and prospects.* Stockholm: Stockholm Environment Institute.

Rijsberman, F. 2006. Water scarcity: Fact or fiction? *Agricultural Water Management* 80: 5–22.

Rogers, P. 2008. Facing the freshwater crisis. *Scientific American* 299: 46–53.

Rogers, P., and Leal, S. 2010. *Running out of Water: The looming crisis and solutions to conserve our most precious resource.* New York: Palgrave Macmillan.

Savenije, H. 2000. Water scarcity indicators: The deception of the numbers. *Physics and Chemistry of the Earth* 13(3): 199–204.

Seckler, D., Amarasinghe, U., Molden, D. J., de Silva, R., and Barker, R. 1998. *World Water Demand and Supply, 1990 to 2025: Scenarios and issues.* IWMI Research Report 19. Colombo: IWMI.

Simon, J. 1980. Resources, population, environment: An oversupply of false bad news. *Science* 208(4451): 1431–1437.

Sullivan, C. A., Meigh, J. R., Giacomello, A. M., Fediw, T., Lawrence, P., Samad, M., Mlote, S., Hutton, C., Allan, J. A., Schulze, R. E., Dlamini, D. J. M., Cosgrove, W. J., Delli Priscoli, J., Gleick, P., Smout, I., Cobbing, J., Calow, R., Hunt, C., Hussain, A., Acreman, M. C., King, J., Malomo, S., Tate, E. L., O'Regan, D., Milner, S., and Steyl, I. 2003. The water poverty index: Development and application at the community scale. *Natural Resources Forum* 27: 189–199.

UN. 2006. *Beyond Scarcity: Power, politics and the global water crisis.* UN Human Development Report. New York. Available online: http://hdr.undp.org/en/media/HDR06-complete.pdf

UNDESA (UN Department of Economic and Social Affairs). 2007. *International Decade for Action 2005–2015: Water for life water scarcity.* Available online: www.un.org/waterforlifedecade/scarcity.shtml

UNESCO (United Nations Educational, Scientific and Cultural Organization). 2006. Water: A shared responsibility. United Nations World Water Assessment Program. Paris: UNESCO.

UNEP. 2008. *Vital Water Graphics: An Overview of the State of the World's Fresh and Marine Waters.* Available online: www.unep.org/dewa/vitalwater/article192.html

UN-Water. 2006. *Coping with Water Scarcity.* UN Water Thematic Initiatives. Available online: www.unwater.org/downloads/waterscarcity.pdf

UN-Water and FAO. 2007. *Coping with Water Scarcity: Challenge of the 21st century.* Available online: www.fao.org/nr/water/docs/escarcity.pdf

Vidal, J. 2012. Water scarcity "could force worldwide vegetarianism." *The Mail and Guardian.* South Africa. Available online: http://mg.co.za/article/2012–08–27-water-scarcity-could-force-worldwide-vegetarianism

WHO. 2003. *Water Sanitation and Health.* Information available online: www.who.int/water_sanitation_health/hygiene/en

World Commission on Environment and Development. 1987. *Our Common Future.* Oxford: Oxford University Press.

World Water Council. 2001. *World Water Vision.* Available online: www.worldwater council.org/fileadmin/wwc/Library/Publications_and_reports/Visions/Commission Report.pdf

# 3 Water Governance

Jonathan Lautze, Sanjiv de Silva, Mark Giordano, and Luke Sanford[1]

## 3.1 Introduction

No one is against good water governance. In fact, with economically viable supply-side options decreasing and demand management tools not always delivering desired results, water governance has emerged as perhaps the most important topic in the international water community in the twenty-first century (Rogers and Hall, 2003; UNDP, 2004; UNESCO, 2006). The 2001 Bonn International Conference on Freshwater, a precursor to the 2002 Johannesburg World Summit on Sustainable Development (Bonn, 2001), identified water governance as the first of three areas of priority action. The second World Water Development report (UNESCO, 2006) highlighted the central role of water governance in improving water resource conditions and boldly stated that "the world water crisis is a crisis of governance—not one of scarcity." The World Bank has also recognized the importance of key governance tenets, such as accountability, in its efforts to reduce the impact of water scarcity (Bucknall, 2007).

Although acknowledgement of and appreciation for water governance's importance is widespread, definitions of the concept can be broad and fuzzy, and inconsistencies in usage and interpretation are common. Castro (2007), for example, points out how UNESCO literature contains a contradiction between i) treating water governance as an instrument to achieve certain goals, and ii) a process that defines the goals. Sehring (2009) asks whether a viable definition of water governance exists and is even possible to create. Tortejada (2006) states that water governance is simply an amalgamation of concepts already in use but under a new, trendy label. Finally, Franks and Cleaver (2007, p. 292) note the "lack of theoretical analysis and debate on core concepts of water governance," and question the very assumption that good water governance leads to good water outcomes.

While this might seem like another semantic issue of development jargon, the definition of water governance has real implications for financial resources and policy as well as actual water resources outcomes. Major donors—including the Swedish International Development Agency (SIDA), the US Agency for International Development (USAID), and the United

Nations Development Programme (UNDP)—have now funded projects and programs focused on improving water governance. Likewise, major educational efforts have been launched to train water professionals in water governance, such as the efforts of the Global Water Partnership and the Arab Water Academy.

To provide a foundation so that such significant investments can hit their mark (i.e., improve water governance), this chapter suggests refining the definition of water governance and the related qualities of good water governance so that they are consistent with the broader notions of governance as understood outside the water sector. The chapter first reviews definitions of governance and water governance (section 3.2). Commonly cited principles of good or effective water governance are then examined (section 3.3). The chapter next examines meanings of water governance and good water governance relative to those of water management and, especially, Integrated Water Resources Management (IWRM) in order to understand some of the underlying sources for confusion surrounding the concepts (section 3.4). Key anomalies and inconsistencies are then highlighted and discussed (section 3.5). Finally, a refined definition of water governance and set of "good" water governance principles are produced as the outcome of this critical re-examination (section 3.6). The chapter's findings reveal fundamental inconsistencies between definitions of water governance and certain principles of good water governance, spurred by a basic conflation of *process* and *outcome*. More broadly, the results provide an improved conceptual basis on which real advances in water governance and water outcomes can potentially be achieved.

## 3.2  Governance and Water Governance

Use of the word "governance" dates back to ancient Greek times, when the term applied to government and simply meant *to steer* (Jessop, 1998). Not surprisingly, definitions and conceptualizations of governance have grown in length and number in recent times, and application of the term has generally broadened to include non-state actors such as civil society, the private sector and non-government organizations (NGOs) rather than simply government (Stoker, 1998; Tropp, 2007). While an exhaustive search for definitions of governance could likely produce hundreds of results, eight definitions sourced from key thinkers on the topic are shown in Table 3.1.

These definitions include three core concepts and exclude one key concept that, as will be shown in later sections, is often included as part of water governance. First, governance is consistently viewed as the *processes* involved in decision making. Second, the processes of decision making take place through *institutions* (including mechanisms, systems, and traditions). Third, the processes and institutions of decision making involve *multiple actors*. Thinking about what governance is through these core aspects also sheds light on what governance is not. That is, governance is the processes

*Table 3.1* Selected definitions of governance

| Definition | Source |
|---|---|
| Governance is a *process* whereby societies or organizations make their important decisions, determine whom they involve in the process and how they render account. | Graham et al. (2003) |
| Since a process is hard to observe, students of governance tend to focus our attention on the governance system or framework upon which the process rests—i.e., the agreements, procedures, conventions, or policies that define who gets power, how decisions are taken, and how accountability is rendered. | |
| The *process* whereby elements in society wield power and authority, and influence and enact policies and decisions concerning public life, and economic and social development. | International Institute of Administrative Sciences (1996) |
| The traditions and institutions by which authority in a country is exercised. This includes the process by which governments are selected, monitored, and replaced; the capacity of the government to *effectively* formulate and implement sound policies; and the respect of citizens and the state for the institutions that govern economic and social interactions among them. | Kaufmann et al. (2005) |
| The concept of governance is viewed by IGS as the sum total of the *institutions* and *processes* by which society orders and conducts its collective or common affairs. | Institute of Governance Studies (2008) |
| The *process* of decision making and the process by which decisions are implemented (or not implemented). | UNESCAP (2009) |
| The exercise of political, economic, and administrative authority to manage a nation's affairs. It is the complex *mechanisms, processes, and institutions* through which citizens and groups articulate their interests, exercise their legal rights and obligations, and mediate their differences. | UNDP (1997) |
| Summary by the Asian Development Bank (ADB) Institute of existing literature on governance: <br><br> • the *processes* by which governments are chosen, monitored, and changed; <br> • the *systems* of interaction between the administration, the legislature, and the judiciary; <br> • the ability of government to create and to implement public policy; <br> • The *mechanisms* by which citizens and groups define their interests and interact with institutions of authority and with each other. | ADB Institute (2005) |
| The manner in which power is exercised through a country's economic, political, and social institutions. | Miller and Ziegler (2006) |

and institutions involved in decision making and *not* the outcomes of that decision making (Rauschmayer et al., 2009).

Governance related to water appears to have first reached the international stage at the World Water Forum in the Hague in 2000, where ministers boldly, though circularly, called for *governing water wisely to ensure good governance* (Rogers and Hall, 2003). Shortly thereafter, in 2001, water governance achieved further prominence when it was identified as the first of three areas of priority action at the International Conference on Freshwater in Bonn, a precursor to the 2002 World Summit on Sustainable Development (Bonn, 2001; Sehring, 2009). Attended by a range of actors including politicians, diplomats, and NGO representatives, the conference helped to spur a prominent role for water governance for years to come. Indeed, numerous international organizations (SIDA, USAID, World Bank, etc.) funded projects explicitly focused on improving water governance in the ensuing years.

Simultaneous with the term's rise to prominence, a number of organizations associated with the burgeoning international water movement began to operationalize the term. Two research and networking bodies were particularly relevant: the Global Water Partnership (GWP) and UN-Water. The GWP was founded in 1996 by the World Bank, UNDP, and SIDA to provide an organizational umbrella for coordinating IWRM activities. Although created with the explicit objective of promoting and implementing IWRM, the GWP also took up the issue of water governance by supporting the development of regional policy papers on the topic, fostering related discussions at workshops and, importantly, developing a definition of the term. UN-Water was launched in 2003 as a coordination mechanism to support member states in their efforts to achieve water-related targets, such as those contained in the Millennium Development Goals (MDGs).[2] UN-Water's flagship program, the World Water Assessment Program (WWAP) monitors freshwater issues in order to provide recommendations, develop case studies, enhance assessment capacity at a national level, and inform the decision-making process (Conca, 2006).

The most prominent definition of water governance was originally produced by the GWP (2002) and subsequently utilized in both GWP and UN-Water WWAP documents. It defines water governance as "the range of political, social, economic and administrative *systems* that are in place to develop and manage water resources, and the delivery of water services, at different levels of society" (GWP, 2002). While not receiving the same level of attention, two other definitions are noteworthy: i) "the different political, social and administrative *mechanisms* that must be in place to develop and manage water resources and the delivery of water services at different levels of society" (GWP-Med, 2001, slide 2), ii) "the political, economic and social *processes and institutions* by which governments, civil society, and the private sector make decisions about how best to use, develop and manage water resources" (UNDP, 2004, p. 10).

These three definitions appear to manifest a marriage of broader concepts of governance with sector-specific concepts of water resources management and service delivery. Governance notions of systems, mechanisms, processes, and institutions on the one hand are combined with water resources development and management on the other. As with general definitions of governance, the presence of multiple actors is again apparent here, often with explicit attention to the fact that actors may exist at different levels or scales.

Despite general consistency with broad concepts of governance, a subtle difference in orientation may nonetheless exist within these water governance definitions. Whereas the focus of UNDP's (2004) definition appears to be on *decision-making processes and institutions* for *how* to use water, GWP definitions treat water governance as *systems and mechanisms* that *are or must be in place* to use water.[3] In other words, whereas UNDP's definition is about processes that decide how water is used, GWP's definitions are more synonymous with a set of institutions that should be in place to use water. The latter definition places less emphasis on dynamic processes of decision making and reflects movement toward treating water governance as simply institutions related to water, perhaps opening the door for a more mechanistic and prescriptive understanding of the term.

## 3.3 Principles for Improving Water Governance

At a practical level, the most important discussion related to water governance centers on identifying principles of good water governance. Principles of good or effective water governance create the important bases for assessing the state of water governance in a given location, and it is through these assessments that opportunities for improvement can be identified. Most development projects in the field are indeed focused on *improving* water governance. Critically important, then, is ensuring that principles for improving water governance are consistent with the concept of water governance, for example the focus is on processes and institutions for decision making rather than pre-decided water resources outcomes.

Somewhat surprisingly, widely recognized qualities of good or effective water governance are in short supply. Perhaps the only semi-authoritative set of principles or criteria was produced by the GWP (Rogers and Hall, 2003) and subsequently adapted and utilized in UN-Water documents (UNESCO, 2003; UNESCO, 2006; UNESCO, 2009). The GWP's 12 principles[4] for effective water governance (Rogers, 2002; Rogers and Hall, 2003) and UN-Water's (UNESCO, 2003; UNESCO, 2006) ten criteria for effective water governance are largely consistent, with differences relating only to grouping of the concepts. Provided here are the GWP's 12 principles for effective water governance:

- open
- transparent
- participative
- accountable
- effective
- coherent
- efficient
- communicative
- equitable
- integrative
- sustainable
- ethical.

This list articulates many of the key elements that one would expect to find, and that would be widely held to be consistent with good or effective govern-ance in general. Nonetheless, given that water governance is essentially the processes and institutions through which decisions are made related to water, it follows that effective water governance is the extent to which qualities that enable effective decision making are actually present in facilitating decision-making processes. Although many of these water governance principles reflect characteristics of a good process (e.g., open, transparent, participative, accountable, communicative, ethical) as one would expect, some of them (e.g., efficient, equitable, integrative, effective, sustainable) are in fact more associated with water resources outcomes.[5] For example, participation and transparency can be reflected in decision making by the level of stakeholder inclusion and the degree of open disclosure of information and decision-making criteria, respectively. By contrast, efficient and sustainable water use comprises outcomes that may result from a water governance process, but outcomes that say little about the strength of the process in and of themselves.

Make no mistake that achieving efficient and sustainable water use is a valid, worthwhile goal. Nonetheless, equating effective governance with effective outcomes implies a far different meaning for water governance than that assigned to it by its definition,[6] and dilutes issues that are core to water governance with those that are extraneous to the concept. If water governance is truly as important as has been widely asserted, it makes sense to understand and assess it on its own terms without introducing water resources outcomes. Not only are outcomes conceptually distinct from gov-ernance processes and hence not part of water governance, notions of uniform good water resources outcomes are practically inconsistent with the variation in values and preferences from one place to the next—variation that gives rise to different perceptions of "good."

Flaws can be seen in the current list when examined from other perspectives as well. For example, there is potential inconsistency between the equity and efficiency principles as well as between participation and sustainability. What if increased equity reduces efficiency? The possibility of such trade-offs have been long documented. Or what if high participation leads to the decision to use water unsustainably? Many countries have in fact consciously decided to use fossil groundwater (and all have decided to use fossil fuels), a quintessentially unsustainable practice. According to principles listed above, is water governance thereby ineffective in almost every country? Additional inconsistencies are also apparent. The key, it appears, is that where potential inconsistencies exist, at least one outcome principle is involved.

## 3.4   Sources of Confusion: Water Management and IWRM

It may in fact be that the outcome–orientation of certain effective water governance principles is triggered by an absorption, inadvertently or not, of water governance into other prominent water sector paradigms (and/or vice versa). Indeed, while water governance is distinct from water management and IWRM,[7] a review of some current water sector documents finds the three concepts frequently intermingled and at times interchanged with little apparent difference in meaning. To understand relationships among these three topics as means to identifying sources of confusion and achieving greater clarity on what water governance *is and is not*, this section reviews and compares meanings of water management and IWRM with water governance.

Water (resources) management has been defined as "the application of structural and nonstructural measures to control natural and man-made water resources systems for beneficial human and environmental purposes" (Grigg, 1996) and "the study, planning, monitoring, and application of quantitative and qualitative control and development techniques for long-term, multiple use of the diverse forms of water resources" (WHO, 2009, n.p.). Whereas water governance is the set of processes and institutions through which management goals are identified, water management is charged with implementing the practical measures to achieve the identified goals. More simply, water management aims to improve outcomes directly, where water governance seeks to define what good outcomes are and align management practices with those goals. Considering *water management* alongside *water governance*, therefore, suggests that water governance provides the framework for deciding on and undertaking management activities. While water governance is at the core of planning activities, governance is relevant even after the shift from planning to implementation. For example, undertaking a practical task such as monitoring groundwater withdrawal can still be affected by elements of water governance such as transparency.

Discussion of water management has grown increasingly moot as use of the term is frequently supplanted by IWRM. IWRM has been defined by GWP (2000, p. 22) as "a process which promotes the coordinated development and management of water, land and related resources, in order to maximize the resultant economic and social welfare in an equitable manner without compromising the sustainability of vital ecosystems" and by USAID (2007, n.p.) as "a participatory planning and implementation process, based on sound science, that brings stakeholders together to determine how to meet society's long-term needs for water and coastal resources while maintaining essential ecological services and economic benefits."[8] While the inclusive nature of IWRM (Jonch-Clausen and Fugi, 2001; Biswas, 2004; Molle, 2008) likely means that water governance, like many other concepts, is subsumed within it, the set of predetermined goals or outcomes associated with IWRM would appear to circumscribe a major role of water governance—that of determining goals. Indeed, how participatory can a planning process be if the goals are predetermined by IWRM constructs?

Incorporation of water governance within the IWRM paradigm may in fact highlight a broader pattern of co-opting water governance into the frameworks of pre-existing organizational mandates, which has a knock-on effect of diluting the conceptual clarity of water governance, undermining its value, confusing practitioners and closing development options. In GWP literature (e.g., Rogers and Hall, 2003), for example, water governance is often treated merely as a tool or prescription to achieve outcomes associated with IWRM. As stated in GWP's formative document on water governance (Rogers and Hall, 2003, p. 16), "water policy *and the process for its formulation* must have as its goal the sustainable development of water resources." Rather than using a water governance process to define a goal, it is a water governance process with a predefined goal. Analogously, UN literature often treats water governance as an instrument to achieve certain goals, such as the MDGs (UNDP, 2004; Castro, 2007) rather than a process to define goals. Interestingly, the two organizations have now joined forces to view water governance in an IWRM framework as critical to achieving MDG targets (UN-Water and GWP, 2007).

The overall effect of including water governance within IWRM is that a potentially distinct identity and, more importantly, role for the concept appears lost. If water governance is important *in itself*, like many of the heralded statements in the first section of this document suggest, the decision-making process for setting water management goals should not be relegated to a foregone conclusion. On the contrary, it is an effective governance process that is first needed to determine which tenets of IWRM, if any, are desirable for a specific location. Moreover, disregarding local conditions, preferences, and values to uniformly apply IWRM principles everywhere actually reflects poor water governance.

## 3.5 Discussion: Identifying Inconsistencies and Resolving Anomalies

Previous sections highlighted how water resources goals or outcomes have been incorporated alongside decision-making processes in conceptualizations of good or effective water governance. Incorporation of these predetermined goals and outcome targets appears to be at least partially a byproduct of a conflation of water governance with IWRM. Incorporation of outcome tenets of IWRM into water governance frameworks, in turn, undermines the ability of governance processes to define major water-related goals since such goals have already been determined. As a result, the practical value of water governance is greatly reduced.

A fundamental conflict appears to exist between the more endowed conceptualization of water governance found in the term's definitions and the more subordinated interpretation of the term evidenced in popular usage by the water community. The primary role for water governance stipulated by the formal definition would be to define water management goals, aligning them with local preferences. Popular perceptions of water governance, by contrast, view the concept more as a tool to achieve better water management, the parameters of which appear to be already set according to international standards (e.g., efficiency, equity, sustainability). A large part of the latter perspective of governance as a tool—absent from the logic underlying the interpretations that follow more strictly from the definition—stems from an often-unquestioned assumption that good water governance actually leads to "good" on-the-ground conditions and "good" outcomes are a reflection of good governance.

Comparing governance processes with water resources and economic outcomes in India and China helps shed light on this presumed relationship. In India, governance processes could be considered relatively open and governance structures relatively inclusive. In China, governance processes are relatively closed according to conventional international metrics, as decisions tend to be made in a top-down fashion that may lack transparency and extensive stakeholder consultation. Applying international standards of good governance, therefore, might suggest that the processes are better in India than in China. However, if one were to examine the efficiency and effectiveness of China's on-the-ground water resources outcomes according to conventional indicators (e.g., water productivity), they would appear equal to or better than those of India. In short, despite its comparatively better governance practices, India generally appears to have worse outcomes.

Another example from the Middle East is also helpful. Here certain oil-rich countries (e.g., Saudi Arabia) may be able to achieve good outcomes while bypassing good governance processes, whereas other countries with a less lucrative natural resources base (e.g., Jordan) may not be able to achieve good outcomes even while applying all good governance processes. In Jordan, for example, water cuts in much of the country are routine as ground and surface water levels decline, despite a relatively inclusive and transparent policy planning process. In Saudi Arabia, by contrast, desalination plants enable more water for cities, rendering water cuts less frequent, yet policy planning processes are not as open and participative. This is not to say that good water governance leads to bad outcomes, but rather to say that water governance is not directly linked to outcomes, and it is certainly possible that absent good governance practices in Jordan the water outcomes there would be worse.

Nonetheless, what seems clear from these examples is that better governance may not lead to better management. That is, the impact of better water governance is evidenced not in better water use according to standard criteria, but rather in management that is more aligned with societal goals. In Jordan and India, one assumes that management goals are more consistent with the preferences of populations in those countries owing to the strength of the process on which those goals were decided. In China and Saudi Arabia, conversely, current governance processes make it relatively difficult to determine the extent to which government goals reflect popular preferences. While improving water governance in these countries could likely help achieve greater consistency between management goals and societal preferences and values, whether or not governance improvements lead to "better" water use according to international IWRM standards is entirely unclear.

Interestingly, if one were to evaluate the two types of countries according to current water governance frameworks that include outcome-based principles (Rogers and Hall, 2003; UNESCO, 2006), evaluation results might look roughly similar despite vastly different situations. If, for example, there is no transparency, participation, or control of corruption in decision making, yet oil wealth allows a country to achieve effective, efficient, and sustainable water resources and economic outcomes, an aggregate "governance" evaluation using conventional metrics could still look respectable. The opposite could also be true. In countries with high participation, transparency, and minimal corruption yet which lack finances for technologies to achieve more efficient, effective, sustainable outcomes, an aggregate "governance" score would be only mediocre despite excellent processes that reflect application of all good process principles.

The last example highlights another important distinction that should be explicitly recognized: that of governance and management. If a wealthy country is able to improve effectiveness, efficiency, and sustainability of water resources outcomes by, for example, investing in water conservation technologies and desalination plants, does that reflect a governance or management improvement, or both? Similarly, if a country utilizes a good forecasting and modeling system to help it to predict and avert a flood, does this improved outcome come as a byproduct of improved governance or improved management? If one interprets governance in the strict process sense as specified by the definition, neither improvement necessarily represents improved governance. To the extent that there is high participation, high transparency, and low corruption in the process of deciding to undertake and implement these developments, the decision-making framework behind such developments could reflect good or bad governance. By contrast, increasing the efficiency, effectiveness, and sustainability of water resources outcomes marks an improvement in water management.

In sum, it appears that IWRM principles have come to influence inter-pretations of water governance and notions of good water governance. IWRM is more prescriptive in nature, having largely predefined outcome goals. Governance, on the other hand, is focused on the processes that are used to decide what the goals are. The greater the predefinition of goals, the more attenuated the role for good governance. And importantly, while good water management or IWRM can produce "good" water outcomes without application of good governance principles (e.g., China or Saudi Arabia in the examples above), it is impossible to know that such outcomes are in fact "good" without a participative and transparent governance process to define a good outcome. Indeed, defining "good" in different locations may be the primary role for water governance, as international definitions of good outcomes do not necessarily coincide with those that are locally generated.

## 3.6 Conclusion: Back to Basics

To review, water governance first hit the international stage in 2000. The Hague Ministerial Declaration in that year boldly called for *governing water wisely to ensure good governance* (Rogers and Hall, 2003). Interestingly, the tautological phrasing of this declaration suggests that there has been confusion in the meaning of water governance from the original usage of the term. The meteoric ascent in importance of the concept evident in the Bonn Declaration the following year, where again interpretation of the term appears murky,[9] ensured a level of prominence for the concept that appeared at odds with general knowledge of its meaning. Given the findings of this chapter, one can easily speculate that its rise to prominence may have been to the detriment of its conceptual utility, as analytical scrutiny to enable a stronger understanding of what the term does and does not mean appears grossly scant.

A related issue concerns how the ambiguity and misunderstanding associated with the term have continued unresolved for more than a decade. Why haven't any of the numerous projects focused on water governance managed to produce a stronger general understanding of the term for practitioners and researchers alike? One possible explanation is that the *chicken may have been counted before it hatched.* Before the concept of water governance had even been defined, it was anointed the first of three priority issues in Bonn, and major international actors soon jumped on the band-wagon. By recognizing water governance as an important issue before it was even clear what *it* was, the impetus to understand the term and allocate it an appropriate level of importance may have been reduced.

This chapter has taken a step in re-evaluating water governance and comparing it with related concepts with which the term is often confounded.

Given the persistent ambiguity but continued importance of the concept in discourse and policy, it is time to return back to basics by evaluating the definition of the concept and the qualities of good water governance. To provide a basis for debate and understanding of water governance, we therefore offer a new definition of term followed by a list of good water governance qualities consistent with the general understanding of the term governance outside the water sector. This definition draws directly on the general definition of governance, and is analogous to the definition of environmental governance (World Resources Institute, 2003, p. 3):

> *Water Governance*: Water Governance is the processes and institutions by which decisions are made that affect water. Water governance does not include practical, technical and routine management functions such as modeling, forecasting, constructing infrastructure and staffing. Water governance does not include water resources outcomes.

*Good Water Governance Qualities*
1. Openness and transparency
2. Broad participation
3. Rule of Law (predictability)
4. Ethical, including integrity (control of corruption)

## Notes

1. This chapter has been adapted from a previously published document: Lautze, J., de Silva, S., Giordano, M., Sanford, L. 2011. Putting the Cart before the Horse: IWRM and water governance. *Natural Resources Forum* 35(1): 1–8.
2. In 2003, UN-Water was endorsed as the new official United Nations mechanism for follow-up of the water-related decisions reached at the 2002 World Summit on Sustainable Development and the Millennium Development Goals.
3. Indeed, absent from GWP definitions (GWP-Med, 2001; GWP, 2002) is explicit use of core governance language such as "processes" and "institutions," and "decision making."
4. The number of these principles ranges from seven in consolidated form (Rogers and Hall, 2003) to 12 when left unconsolidated (Rogers, 2002).
5. We appreciate here that several of these principles can be treated as an element of process or outcome. The way the term is used in the report leads one to believe the authors were applying them as elements of an outcome.
6. The report in which these are listed even categorizes some of these qualities separately in a "performance" category.
7. Indeed, if there is no distinction, there is no need for a new term.
8. USAID (2007) continued to explain that "IWRM helps to protect the world's environment, foster economic growth and sustainable agricultural development, promote democratic participation in governance, and improve human health."
9. Indeed, while the list of 12 specific actions under the rubric of water governance include some attributes commonly linked with governance (e.g., participation, gender equity, fight corruption), many core water resources management functions are also present; e.g., "improve water management," "protect water quality and ecosystems," "manage risks to cope with variability and climate change."

# References

Asian Development Bank (ADB) Institute. 2005. Governance in Indonesia: Some Comments. Discussion Paper No: 38. Tokyo: Asian Development Bank, p. 22.

Biswas, A. 2004. Integrated water resources management: A reassessment. *Water International* 29(2): 248–256.

Bonn. 2001. International Conference on Freshwater. Bonn, Germany, 3–7 December. Available online: www.water-2001.de

Bucknall, J. 2007. *Making the Most of Scarcity: Accountability for better water management in the Middle East and North Africa*. Washington DC, World Bank, p. 235.

Castro, J. 2007. Water governance in the twenty-first century. *Ambiente & Sociedade* 10(2): 97–118.

Conca, K. 2006. *Governing Water: Contentious transnational politics and global institutions building*. Cambridge, MA and London: The MIT Press.

Franks, T., and Cleaver, F. 2007. Water governance and poverty: A framework for analysis. *Progress in Development Studies* 7(4): 291–306.

Global Water Partnership (GWP). 2000. *Integrated Water Resources Management*. TAC Background Paper No 4. Stockholm: Global Water Partnership.

Global Water Partnership (GWP). 2002. *Introducing Effective Water Governance*. GWP Technical Paper.

Graham, J., Amos, B., and Plumptre, T. 2003. *Principles for Good Governance in the 21st Century*. Policy Brief No.15. Institute On Governance. Ontario, Canada, p. 6.

Grigg, N. 1996. *Water Resources Management: Principles, regulations and cases*. New York: McGraw Hill.

GWP-Med. 2001. *Presentation on Effective Water Governance*. Workshop of the GWP-Med Sub-Regional Working Groups of North Africa and Middle East on Effective Water Governance. Marriott Hotel. Cairo, 20–21 December.

Institute of Governance Studies. 2008. *The State of Governance in Bangladesh in 2008: Confrontation, competition, accountability*. Dhaka, Bangladesh: Institute of Governance Studies, BRAC University.

International Institute of Administrative Sciences. 1996. *The Governance Working Group*. Available online: www.gdrc.org/u-gov/work-def.html (accessed in March 2009).

Jessop, B. 1998. The rise of governance and the risks of failure: The case of economic development. *International Social Science Journal* 50(155): 303–316.

Jonch-Clausen, T., and Fugi, J. 2001. Firming up the conceptual basis of integrated water resources management. *International Journal of Water Resources Development* 17(4): 501–510.

Kaufman, D., Kraay, A., and Mastruzzi, M. 2005. Governance Matters IV: Governance Indicators for 1996–2004. Washington, DC: World Bank.

Miller, U., and Ziegler, S. 2006. *Making PRSP Inclusive*. Munich: Handicap International, Christoffel-blindenmission.

Molle, F. 2008. Nirvana concepts, narratives and policy models: Insight from the water sector. *Water Alternatives* 1(1): 131–156.

Rauschmayer, F., Berghofer, A., Omann, I., and Zikos, D. 2009. Examining processes and/or outcomes? Evaluating concepts in European governance of natural resources. *Environmental Policy and Governance* 19: 159–173.

Rogers, P. 2002. *Water Governance in Latin America and the Caribbean.* Washington, DC: Inter-American Development Bank, Sustainable Development Department, Environment Division.

Rogers, P., and Hall, A. 2003. *Effective Water Governance.* TEC Background Paper 7. Stockholm: Global Water Partnership.

Sehring, J. 2009. *The Politics of Water Institutional Reform in Neopatrimonial States: A comparative analysis of Kyrgyzstan and Tajikistan.* New York: Springer.

Stoker, G. 1998. Governance as theory: Five propositions. *International Social Science Journal* 50: 17–28.

Tortejada, C. 2006. *Water Governance, What Does It Mean?* Mexico: Third World Centre for Water Management.

Tropp, H. 2007. Water governance: Trends and needs for new capacity development. *Water Policy* 9, Supplement 2: 19–30.

United Nations Development Programme (UNDP). 1997. *Governance for Sustainable Human Development.* A UNDP Policy Document. New York: UNDP.

United Nations Development Programme (UNDP). 2004. *Water Governance for Poverty Reduction Key Issues and the UNDP Response to the Millennium Development Goals.* Water Governance Programme Bureau for Development Policy. New York: UNDP.

United Nations Economic and Social Commission for Asia and the Pacific (UNESCAP). 2009. *What is Good Governance?* New York: UNESCAP.

United Nations Educational, Scientific and Cultural Organization (UNESCO). 2003. *Water for People, Water for Life: A joint report by the twenty-three UN agencies concerned with freshwater.* Volume 1 of The United Nations World Water Development report. UN World Water Assessment Programme. Paris: UNESCO, p. 576.

United Nations Educational, Scientific and Cultural Organization (UNESCO). 2006. Water Development Report 2: *Water: A shared responsibility.* United Nations World Water Assessment Program. Paris: UNESCO.

United Nations Educational, Scientific and Cultural Organization (UNESCO). 2009. *Water in a Changing World.* United Nations World Water Assessment Program. Paris: UNESCO.

UN-Water and Global Water Partnership. 2007. *UN-Water and Global Water Partnership Roadmapping for Integrated Water Management Processes.* Copenhagen Initiative on Water and Development. Available online: www.ucc-water.org

United States Agency for International Development (USAID). 2007. *What Is Integrated Water Resources Management?* Available online: www.usaid.gov/our_work/environment/water/what_is_iwrm.html (accessed in March 2009).

World Health Organisation (WHO). 2009. Located on Waterwiki.net at http://waterwiki.net/index.php/Water_management (accessed in March 2009).

World Resources Institute. 2003. *World Resources: 2002–2004. Decisions for the Earth: Balance, voice and power.* Washington, DC: World Resources Institute.

# 4 Water Security

*Jonathan Lautze and Herath Manthrithilake*[1]

## 4.1 Introduction

Water security has come to assume an increasingly prominent position in the international water and development community in recent years. Staff at the World Bank have explained that water security is critical for growth and development (Grey and Sadoff, 2007; Grey and Connors, 2009). The importance of water security for the sustainable development of countries such as China has been recognized nationally (Chen, 2004; Cheng et al., 2004; Liu et al., 2007). Water security has been at the heart of high profile negotiations on a Cooperative Framework Agreement in the Nile Basin (WaterLink, 2010). Finally, academia (Briscoe, 2009; University of East Anglia, 2009; Sinha, 2009; Tarlok and Wouters, 2009; Vorosmarty et al., 2010; Zeitoun, 2011; Cook and Bakker, 2012; Lankford et al., 2013) and other development actors (FAO, 2000; Swaminathan, 2001; Asian Development Bank, 2007; Biswas and Seetharam, 2008; Asia Society, 2009) have also placed prominent emphasis on the concept.[2]

Despite the elevated status that the term has increasingly acquired in policy documents and development discourse, the concept of water security remains largely unquantified.[3] While there may be advantages to leaving the concept as a qualitative theoretical ideal, there are simultaneously several benefits to translating water security into numerical terms. First, it can encourage clarity and common understanding of a concept around which there currently exists substantial ambiguity. Second, it can help to foster discussion and debate on scales and thresholds for evaluating the presence, absence, or degree of water security. Third, it can help to assess the extent to which the concept is really being achieved on the ground in different locations.

This chapter devises an index that quantifies water security at a country level in order to encourage a more concrete understanding of the term. An initial section (section 4.2) reviews definitions of water security and identifies five components that provide a conceptual framework for assessment: basic needs, agricultural production, the environment, risk management, and independence. The conceptual framework is then translated into a set of numerical indicators (section 4.3), which are populated with data from 46 countries in the Asia-Pacific region to generate a set of results (section 4.4). The Asia-Pacific was selected because of its great diversity of water resources conditions and economic development levels, and owing to its degree of available data. Finally, key issues revealed through undertaking this approach are examined (section 4.5), and the viability of the approach as well as the added value of water security as a concept are discussed (section 4.6).

## 4.2   Conceptual Framework

As water security is a fairly new concept, definitions of the term appear to be evolving. Reviewing four key definitions of the term suggests that the meaning of water security has grown somewhat more expansive since its initial use, to include more explicit focus on agriculture and food production, adverse impacts of water, and national security. The Global Water Partnership (2000, p. 12) defined water security simply as an overarching goal where "every person has access to enough safe water at affordable cost to lead a clean, healthy and productive life, while ensuring the environment is protected and enhanced." Swaminathan (2001, p. 35) then stated that water security "involves the availability of water in adequate quantity and quality in perpetuity to meet domestic, agricultural, industrial and ecosystem needs." Cheng et al. (2004) subsequently defined water security to include access to safe water at affordable cost to enable healthy living and food production, while ensuring the water environment is protected and water-related disasters such as droughts and floods are prevented. Finally, Grey and Sadoff's (2007) more recent definition of water security is focused on "the availability of an acceptable quantity and quality of water for health, livelihoods, ecosystems and production, coupled with an acceptable level of water-related risks to people, environments and economies."[4]

Despite some differences, these definitions have several common strands. A first common strand is a focus on access to potable water for basic human needs or domestic use. A second relates to provision of water for productive activities—presumably production of agriculture, food, and industrial goods, as specified in some definitions. A third is the focus on environmental conservation or protection. A fourth strand, common at least to the latter two definitions, is prevention of water-related disasters. A final element worth noting relates to Grey and Sadoff's (2007) broader treatment of risk, which strongly suggests inclusion of issues related to water for national security or independence.

Based on the four common strands and final element specific to Grey and Sadoff (2007), a conceptual framework is hereby proposed that contains five components: basic needs, agricultural production, the environment, risk management, and independence. It should be noted that the focus of the second component was confined to agricultural production, which encompasses food production yet excludes other areas that may plausibly be subsumed within this component such as industry and energy. Focus was confined to agriculture in the second component because agriculture is the largest productive use of water, and because water was considered either too extraneous to outcomes in other areas, non-essential, or establishment of key relationships were too data constrained. With industry, for example, water is but one input among many, and different levels of industrial output are likely most associated with factors other than different levels of water security. As for energy, while some countries rely on hydropower as a critical source of energy, other countries satisfy all their energy requirements without making use of hydropower.[5] Gauging water security related to hydropower in a cross-country fashion, therefore, is severely constrained by the non-essentiality of hydropower for energy production. Finally, while there may be a more direct connection in the case of water for cooling after electricity generation, there was insufficient national-level data on water for cooling so it was not considered.[6]

Importantly, consideration of these five components can be treated as important to *enabling* many of the outcomes linked to water security, such as adequate food consumption, healthy people, economic development, and environmental conservation. However, achieving security in these areas is a function of much more than water security. For example, while water security can imply that economies are buffered from droughts and floods, this does not mean that economies will be resilient from other shocks such as those related to global financial crises. Similarly, while water security implies sufficient agricultural production to feed a community or country, the selection of crops that satisfy nutritional needs, and the distribution and provision of those crops in a time-appropriate manner may not fall within the parameters of water security—this is food security. As such, water security can be considered but one contributor to the security of other areas such as food and environment. Ultimate security in these areas, however, relies on factors over and above those specific to water security.

## 4.3 Methods

To assess water security for basic needs, agricultural production, the environment, risk management, and independence, data were utilized from a combination of recent sources (e.g., FAO AQUASTAT, 2007; WHO, 2009; World Resources Institute, 2009). Methods used to assess water security in each of the five components of the framework are discussed below and summarized in Table 4.1. A quintile-based approach was utilized in each component, whereby countries were ranked according to their performance, divided into five quintiles that were approximately equal in size, and assigned a score depending on the quintile into which they fell.

## A. Water Security for Basic Needs

To assess the degree to which countries have achieved water security for the basic needs of their populations, we utilized data from the World Health Organization (WHO, 2009) on percentage of populations with sustainable access (within 1 km) to an improved water source (household connections, public standpipes, boreholes, protected dug wells, protected springs, and rainwater collection). Results for countries were ranked according to the proportion of their population with sustainable access to an improved water source and divided into five groups of roughly equal size. A score between 1 and 5 was assigned to each group: 5 indicates a greater proportion of a country's population has sustainable access to an improved water source, and 1 indicates a smaller proportion has sustainable access to an improved water source.

## B. Water Security for Agricultural Production

The degree to which water security for agricultural production is achieved in a country was treated as a composite of two sub-indicators: i) water availability per capita and ii) water withdrawal per capita. Data for both sub-indicators were obtained from FAO AQUASTAT (2007). Water availability per capita (i.e., renewable water resources/population) provides an indication of total water available for agricultural production. It is particularly relevant to rainfed agriculture in a country, but it also provides an indication of the potential for irrigation. Given that greater water availability enables more rainfed agriculture and greater potential for irrigation, greater water availability per capita can be associated with greater water security for agricultural production. Water withdrawal per capita provides an indication of how much control a country possesses of its water resources. Given that agriculture is the primary user of water in virtually every country, accounting for more than 80 percent of water use in Asia (FAO AQUASTAT, 2007), greater control of water can be associated with greater water security for agricultural production.

For each of the two sub-indicators (water availability per capita and water withdrawal per capita), countries were ranked and divided into five groups. A score between 1 and 5 was then assigned to each group. For the first sub-indicator, a score of 5 reflects greater water availability per capita, and a score of 1 indicates less water availability per capita. For the second sub-indicator, a score of 5 indicates greater water withdrawal per capita and a score of 1 indicates less water withdrawal per capita. Results in each of the two sub-components were then averaged. Therefore 5 represents greater water security for agricultural production in a country, and 1 represents less water security for agricultural production in a country.

Table 4.1 Calculating water security

**Overall Water Security = A + B + C + D + E**

| Component | Definition | Scoring system | | Source |
|---|---|---|---|---|
| A = Basic Household Needs | Percentage of population with sustainable access to an improved water source | High percentage of population with access to improved water source = 5 to low percentage of population with access to improved water source = 1 | | WHO, 2009 |
| B = Food Production | The extent to which water is available and harnessed for agricultural production | Water security for agricultural production = (a+b)/2 | | FAO AQUASTAT, 2007 |
| | | a. Water availability (RWR/population) | From low availability = 1 to high availability = 5 | |
| | | b. Water use (Withdrawal/population) | From low withdrawal = 1 to high withdrawal = 5 | |
| C = Environmental Flows | Percentage of Renewable Water Resources (RWR) available in excess of environmental water requirement (EWR). That is, [RWR − (environmental water requirement + withdrawn water)]/RWR. | High percentage above EWR = 5 to low percentage above EWR = 1 | | converted from Smakhtin et al., 2004 |
| D = Risk Management | Risk Management measures the extent to which countries are buffered from the effects of rainfall variability through large dam storage | Risk Management = (a+b)/2 | | Mitchell et al., 2002; ICOLD, 2003; FAO AQUASTAT, 2007 |
| | | a. Inter-annual CV | From low CV = 5 to high CV = 1 | |
| | | b. Storage | From high storage = 5 to low storage = 1 | |
| E = Independence | Independence measures the extent to which countries water and food supplies are safe and secure from external changes or shocks | From low dependence on external waters = 5 to high dependence = 1 | | WRI, 2009 |

## C. Water Security for the Environment

The degree to which water security for the environment is achieved in a country was considered to be the extent to which environmental water requirements are satisfied. Clearly, achieving sufficient quantities of water for environmental needs captures only part of the picture, as it is also important that water for the environment be of appropriate quality. Nonetheless, since country-level data on water quality were not widely available, focus was devoted solely to water quantity.

To assess country-level water security for environmental flows, we determined the percentage of un-withdrawn water in excess of the environmental water requirement. To calculate this percentage, we subtracted the amount of water withdrawn and the environmental water requirement from a country's renewable water resources (converted from Smakhtin et al., 2004). The remaining amount was then divided by a country's renewable water resources (RWR).[7] Countries were ranked, divided into five groups, and a 1 through 5 score was applied to each group: 5 indicates a greater proportion of water available in excess of the environmental water requirements and 1 indicates a smaller proportion.

## D. Water Security for Risk Management

Recognizing that many essential activities in countries are vulnerable to fluctuations in rainfall and that storing water constitutes a viable method to mitigate the effects of these fluctuations, water security for risk management was considered to be the extent to which water storage capacities are in place to offset a country's level of inter-annual rainfall variability. This indicator contains two sub-components. A first sub-component consists of the percent of renewable water resources stored in each country, calculated by dividing the storage capacity in large dam reservoirs (International Commission on Large Dams, 2003) by the country's renewable water resources (FAO AQUASTAT, 2007). Large dams are admittedly but one storage option, as there are indeed other ways to store water such as in the ground, soil, and behind small dams (IWMI, 2009; Taylor, 2009). Nonetheless, accessible data across countries are only available for water storage behind large dams. Countries were stratified into five groups depending on the percent of their renewable water resources that they store, with higher storage levels scoring greater than lower storage levels.

The second sub-indicator focused on inter-annual rainfall variability, for which we utilized country-level data on the inter-annual rainfall coefficient of variation (Mitchell et al., 2002). Countries were divided into five groups according to the degree of inter-annual rainfall variability, with lower rainfall variability scoring greater than higher rainfall variability. To develop an aggregate score for risk management, each country's scores in the two sub-components were averaged. A 1 through 5 scale was utilized. Scores toward

5 indicate greater water security for risk management (i.e., more storage, less variability). Scores toward 1 indicate less water security for risk management (i.e., less storage, more variability).

### E. Water Security for Independence

Recognizing that a country's national security is tied to the degree to which it is capable of satisfying its own water needs through internal means, water security for independence was considered to be the proportion of a country's water resources generated internally. To determine the proportion of water originating inside a country, we utilized the dependency ratio (World Resources Institute, 2009), an indicator of the proportion of a country's water resources that are generated internally. Countries were categorized on a 1 through 5 scale such that a score of 5 indicates that a country is largely reliant on its own water resources and 1 indicates a heavy reliance on external waters.[8]

### Overall Water Security Index

To generate a score for overall water security, results for each of the five components were summed, producing a 25-point scale (Table 4.1). Just as 5-point scales indicate the degree of water security achieved in individual components, the broader score on a 25-point scale indicates the degree of overall water security in a particular country. In Figure 4.1, therefore, scores for each of the components is on a 5-point scale, and the overall maximum that can be achieved by a country in all five components is 25 points. A higher score indicates greater water security, and a lower score indicates the opposite.

## 4.4 Results

Comparing the strength of overall water security scores across countries reveals substantial dispersion, with scores ranging from less than 10 to greater than 20 (Figure 4.1). Noticeably, even in those countries that appear quite water secure, there still exist weak spots. Despite Australia's overall high level of water security, for example, the specific component of risk management appears only mediocre. Similarly, Japan appears limited by its poor score in water security for the environment, and Malaysia could do with improvement in water security for the environment and independence. Conversely, in water insecure countries such as Cambodia and Afghanistan, weak spots are apparent in at least four of the five components in the water security framework.

Some results for overall water security defy perceptions that water security is tied to economic development (Figure 4.2). For example, it was somewhat surprising to find countries such as Myanmar, Bhutan, and the Kyrgyz

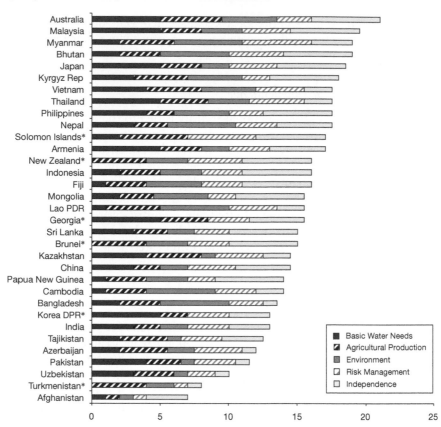

*Figure 4.1* Water security in the Asia-Pacific, ordered from greatest to least water secure.

*Indicates that data are available for only four rather than all five components. Countries with data for less than four components are not displayed.

Source: authors' calculations.

Republic among those with the greatest level of overall water security. Nonetheless, these countries are all quite water endowed, with much of their water resources generated internally, and with less alteration to the environment than many other countries. Hence their scores in certain components may have been sufficient to buoy these countries' overall water security scores.

A review of scores in individual water security components reveals results that could be largely predicted based on levels of economic development, yet which provide occasional surprises (Figure 4.3). Countries scoring highly in water security for basic water needs (e.g., Australia, Georgia, Malaysia), for example, are among the more developed in the Asia Pacific or are former Soviet Republics. Countries scoring low in basic water needs

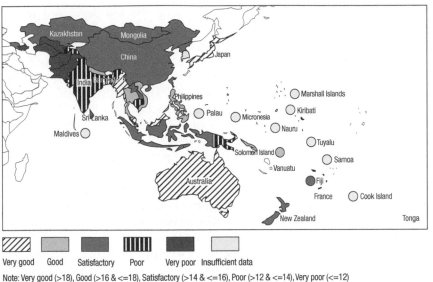

Very good   Good   Satisfactory   Poor   Very poor   Insufficient data

Note: Very good (>18), Good (>16 & <=18), Satisfactory (>14 & <=16), Poor (>12 & <=14), Very poor (<=12)

*Figure 4.2* Overall water security.
Source: authors' calculations.

Very good   Good   Satisfactory   Poor   Very poor   Insufficient data

Note: Very good (>4), Good (>3 & <=4), Satisfactory (>2 & <=3), Poor (>1 & <=2), Very poor (<=1)

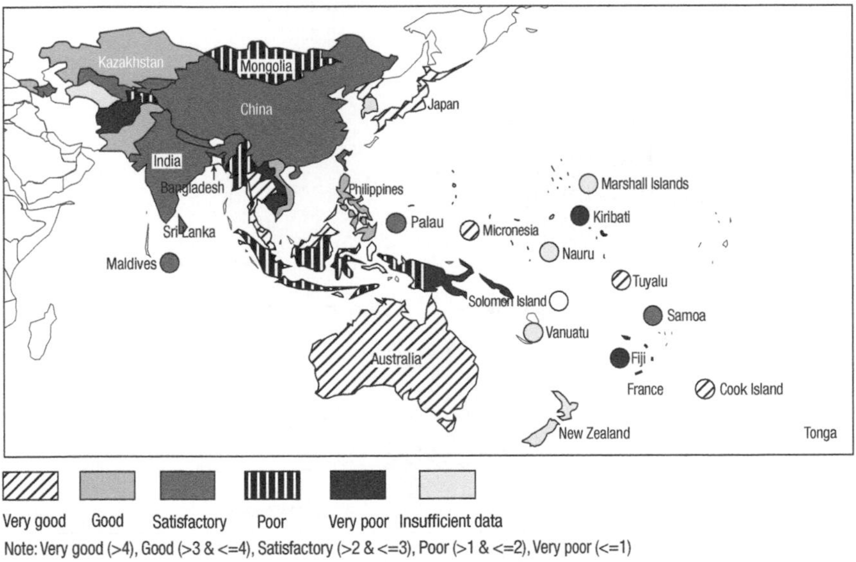

*Figure 4.3* Water security for basic needs.
Source: authors' calculations.

(e.g., Afghanistan, Cambodia, Fiji) are among the less developed in the region or are small island states. Overall, the results of this component yield few surprises and could be said to be largely aligned with expectations.

Results related to water security for agricultural production were somewhat less aligned with levels of economic development (Figure 4.4). Kazakhstan, the Kyrgyz Republic, Lao PDR, Myanmar, Turkmenistan, and Vietnam, for example, were among countries that scored fairly high. The Korean Republic and Singapore, by contrast, scored fairly low. In the first group of countries, the results reflect the fact that mean quantities of renewable water resources per person, and levels of withdrawal per person, are relatively high. The latter group of countries has low per capita water availability and low water withdrawal per person. These findings may highlight the potential for greater agricultural production in the former group of countries.

Results related to water security for the environment (Figure 4.5) indicate Southeast Asian countries to be relatively strong and Central Asian countries to be somewhat weak. Countries scoring highly (e.g., Bangladesh, Cambodia, Lao PDR, Myanmar, Nepal) are concentrated in conditions of somewhat low levels of water resources development. Countries scoring lower (e.g., Kazakhstan, Pakistan, Tajikistan, Uzbekistan) are concentrated in countries with higher levels of water resources development.

Many of the countries scoring highly in water security for risk management (e.g., Bhutan, New Zealand, Singapore) would likely be predicted to be effectively managing water-related risk (Figure 4.6). Other countries such

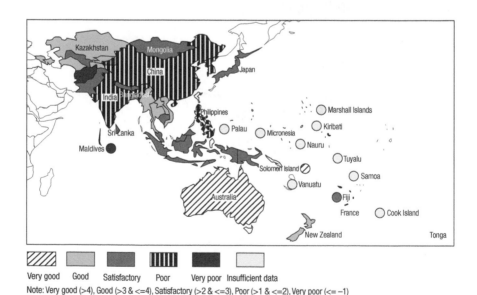

*Figure 4.4* Water security for agricultural production.

Source: authors' calculations.

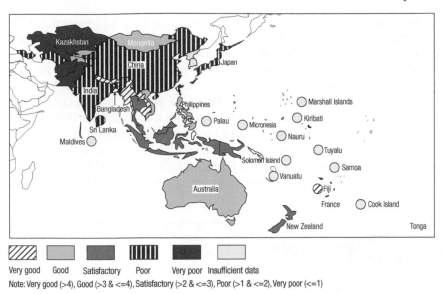

*Figure 4.5* Water security for the environment.

Source: authors' calculations.

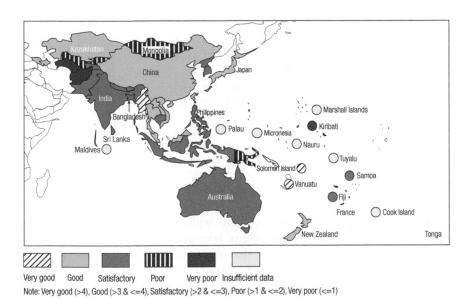

*Figure 4.6* Water security for risk management.

Source: authors' calculations.

as Myanmar, Solomon Islands, and Vanuatu might be less expected to be effectively managing risk. Scores in some countries may nonetheless appear deceivingly high due to the impacts of one storage infrastructure on a small water resources base.[9] Countries scoring low in water security for risk management are mainly those with lower levels of economic development.

A review of the results for the final component in the assessment framework, water security for independence, yields few surprises (Figure 4.7). Countries that are islands and located in upstream portions of basins fare better than downstream nations. Countries such as Australia are more water secure by virtue of their geographic position, for example, while other countries such as Bangladesh are fairly insecure due to their heavy reliance on inflows from upstream countries.

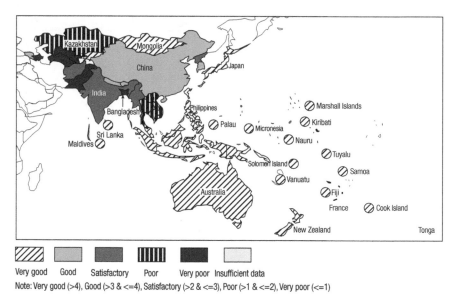

Very good    Good    Satisfactory    Poor    Very poor    Insufficient data

Note: Very good (>4), Good (>3 & <=4), Satisfactory (>2 & <=3), Poor (>1 & <=2), Very poor (<=1)

*Figure 4.7* Water security for independence.

Source: authors' calculations.

## 4.5  Discussion

The chapter identified five key components of water security and translated them into numerical indicators that were applied across the countries of the Asia-Pacific. While the results might spur few surprises if presented in countries due to local knowledge that may already exist on water sector strengths and weaknesses, a primary benefit of applying a water security framework such as this is to understand how water secure countries are in relation to one another. A secondary benefit, if the framework is reapplied in the future, is monitoring the rate and direction of change in water security to enable comparison over time.

An important goal of this chapter was identifying some of the key issues inherent in assessing water security in order to spur more concrete discussion on what the concept truly means. One fundamental issue raised by the methods employed relates to assessment of relative v. absolute water security. As apparent from the methods section above, the approach utilized in this chapter assessed *relative* water security. Either approach has advantages and limitations. Assessment of *relative* water security allows for the reality that there is not necessarily an ideal state of water security, and that notions of good water security will be in constant evolution and implicitly affected by known reference points (e.g., on-the-ground conditions in countries). Treating notions of good water security as relative, however, fails to reflect the potential that the best levels of water security on-the-ground may still be poor. By contrast, while evaluation of *absolute* water security enables assessment of countries according to more standardized thresholds, identification of such thresholds would be no easy task, and might be derived from practical country-level experiences anyway. Further, use of absolute indicators could imply existence of an ideal state of water security, which is debatable.

Whichever the case, another issue raised by the approach utilized in this chapter relates to the scale at which water security is assessed. While country-level assessment has advantages in particular for international donors who typically transfer development funds to national governments, evaluation of water security conditions at a country level is inconsistent with the fact that water management is often conducted at a basin level. In some countries, all basins fall within national borders, such that a country-level water security score can be considered an aggregation of water security in specific basins. However, many other countries contain basins that cross borders, which may confound results determined at a country level. A particular country may have insufficient storage on its own territory to mitigate the effects of rainfall variability, for example, but may be able to rely on the storage capacities of an upstream neighbor. Similarly, while a country may generate too little water internally to satisfy its national security needs, inflow from an upstream country may be sufficiently assured through an international agreement.

In light of the confounding nature of transboundary waters to country-level evaluation of water security, one way to improve the assessment framework is to include the existence and functions of transboundary water agreements. If a country relies on external waters but such waters are assured through a treaty, for example, that country is clearly in a more water secure position than an analogous country without a treaty (Sadoff et al., 2008). To capture this nuance, the amount of water assured by a provision in an international treaty could be added to that which a country generates internally. Although there are cases when treaty provisions are not honored, implying water assured through a treaty is not as secure as that produced internally, consideration of transboundary treaties would nonetheless help to reflect the reality that water management is often undertaken at the basin level, even in the context of transboundary waters.

A third issue revealed by our approach relates to the conceptualization of water security for agricultural production. There was a temptation related to this component to make use of conventional indicators in agricultural water management, such as water productivity and related sub-indicators of efficiency or yield per unit of evapotranspiration (ET). The approach utilized, however, measured water availability and use that *enable* agriculture and food production. While improving water productivity is clearly a way to increase agricultural production, it is simply one means to improving agricultural production, and may not be essential. In areas of economic water scarcity, for example, greater storage may be needed more urgently than improved productivity (Molden et al., 2010).

A fourth issue raised by our approach relates the inclusion of water storage behind large dams and exclusion of other forms of storage such as groundwater and soil moisture. While obviously our analysis would have been strengthened through inclusion of all forms of storage, national data for water stored in the ground and soil simply do not exist at this point. This data constraint may have biased the analysis in favor of those countries that focus on large dam storage, and have somewhat underestimated the water security in countries that make more effective use of groundwater and soil water to buffer themselves from the effects of rainfall variability.

A final point relates to the aggregation of the five components into an overall water security index. While it is possible to perform well in all components of the assessment framework, it must be acknowledged that the performing well in one component of water security may adversely affect performance of other components, and vice versa. In particular, achieving higher levels of water security for agricultural production and risk management may require higher levels of water storage and use, which may decrease water available for the environment. Conversely, ensuring ample water for the environment may constrain scores in water security for agricultural production and risk management.

## 4.6  Conclusion

The development and application of the approach utilized in this chapter has helped clarify the notion of water security, and prompts at least two overarching suggestions for understanding the meaning and practical utility of the concept. A first suggestion for reaching more common understanding of the concept is to move beyond qualitative definitions to make a list, or finite set of criteria, on which water security is determined and evaluated, as proposed in this chapter. While the criteria utilized in this chapter may not be perfect, it is believed they mark a valuable step toward arriving at a clear meaning of the concept. A second suggestion is to clearly distinguish between means and ends. Analogous in some ways to the need to disentangle the common conflation of processes and outcomes in the context of water governance (see Chapter 3), interpretations of water security could benefit from clear focus on the end of water security—not the means to water security, and not the ends beyond water security.

Interestingly, given the current ambiguity associated with the concept, it seems somewhat ironic that so much importance is attached to it. In high profile negotiations over a comprehensive agreement on Nile Basin waters, for example, water security is considered the paramount issue on which negotiations have been stalled for multiple years (WaterLink, 2010). Yet why should governments agree to such a concept if a set of its key elements have not been clearly defined, and hold potential to undermine their positions if more exhaustively outlined at a subsequent point? Or, conversely, what is the point to agreeing to a concept that can ultimately be interpreted in multiple ways in the future?

In terms of the issue posed at the outset of this document about the added value of introducing the concept of water security, the results are mixed. While focusing on five priority issues related to water management is important, the benefits of bundling these five issues under the umbrella of a new paradigm are not immediately apparent. On the contrary, with so many other new concepts related to water permeating discourse (e.g., IWRM, water governance, hydropolitics), there may be confusion, skepticism, and even fatigue associated with the introduction of another new term that is not concretely defined yet which is supposed to comprise a panacea for water managers.

There has indeed been a steady flow of new terms in the water management community in the last decade that have not been matched by a steady flow of clear meanings, which has engendered reactions of *eye-rolling* at the introduction of additional terms such as water security. Nonetheless, while the lack of clear widely accepted meanings would appear to be to the terms' detriment, one simultaneously wonders whether there may be benefits to leaving the terms vague. If water security is de-shrouded to reveal that it is simply a package of five criteria that have already been utilized for decades, for example, the need for packaging may be questioned and the topic may lose some of its allure.

From a more practical perspective, the need to aggregate the five components in the assessment framework into an overall score is questionable, for at least two reasons. First, presentation of overall scores for countries above typically triggers—almost immediately—interest in identifying the specific areas that explain such overall scores. Second, related, presentation of the overall water security scores provides little direct guidance to countries, given all the information compressed into one value. Presentation of results at a component-level, by comparison, provides indications of the factors explaining water security performance, which in turn provide a basis for recommendations for improving conditions.

In sum, the approach utilized in this chapter constitutes an initial effort to assess the central components of water security and identify some of the major issues in undertaking such an exercise. There is clear value in prioritizing critical areas in water management, and in evaluating and comparing them. Insofar as water security identifies the priority areas in water management, therefore, there is added value in the concept. Nonetheless, the need to package the set of priority areas under a new label is not clear, and indeed appears a mixed blessing. On the one hand, use of a well-chosen heading can spur—and has spurred—interest in an important set of key areas for water management. On the other hand, use of such a heading may engender confusion and inflated expectations of the concepts subsumed beneath it.

## Notes

1. This chapter is an updated version of a previously published document: Lautze, J., and Manthrithilake, H. 2012. Water security: Old concepts, new package, what value? *Natural Resources Forum* 36(2): 76–87.
2. We acknowledge that some of these documents feature the language of water security prominently yet use the term quite loosely.
3. Only Maplecroft (2011), a private sector UK-based consulting firm, has attempted to quantify "water security risk" as a guide for private sector investment in countries. Their methods are somewhat untransparent, however, and it is not clear that they have released formal publications on their work. Vorosmarty et al. (2010) have assessed threats to water security rather than water security per se.
4. As this book goes to press, the United Nations University (2013) has also offered a "working definition" which appears largely a composite of those presented here:

   > The capacity of a population to safeguard sustainable access to adequate quantities of and acceptable quality water for sustaining livelihoods, human well-being, and socio-economic development, for ensuring protection against water-borne pollution and water-related disasters, and for preserving ecosystems in a climate of peace and political stability.

5. One option to consider water security for energy is to stratify countries according to the degree to which hydropower contributes to their energy production. In the subset of countries in which hydropower satisfies a major portion of energy requirements, a supplemental indicator could be used to gauge water security for energy.
6. Quantities of biofuel and desalinated water production are two other areas that may be considered in future analysis. At present, however, their use would appear too limited in most countries to justify incorporation into an assessment framework.
7. In other words, percent in excess of environmental water requirement = [RWR – (environmental water requirement + withdrawn water)]/RWR.
8. Judgment was used to allocate scores of high independence (i.e., 5) to several island countries for which data in this component were unavailable.
9. In the case of Myanmar, limited data in the component of risk management may have worked to elevate the overall score of the country as well.

# References

Asia Society. 2009. *Leadership Group on Water Security in Asia.* Progress Report (June). Asia Pacific Water Forum. Available online: www.apwf.org/archive/documents/5th_GC/3–17_Asia_Society_concrete_initiative_report.pdf

Asian Development Bank (ADB). 2007. *Asia Water Development Outlook: Achieving water security for Asia.* Available online: www.adb.org/Documents/Books/AWDO/2007/awdo.pdf

Biswas, A., and Seetharam, K. 2008. Achieving water security for Asia, *International Journal of Water Resources Development* 24:1: 145–176.

Briscoe, J. 2009. *Harvard Water Initiative: Science technology and policy for water security.* Available online: www.johnbriscoe.seas.harvard.edu/research/Harvard%20University%20Water%20Security%20Initiative%2020090609.pdf

Chen, S. 2004. Differentiating and analyzing the concept of water security. *China Water Resources* 17: 13.

Cheng, J., Yang, X., Wei, C., and Zhao, W. 2004. Discussing water security. *China Water Resources* 1: 21–23.

Cook, C., and Bakker, K. 2012. Water security: Debating an emerging paradigm. *Global Environmental Change* 22: 94–102.

FAO AQUASTAT. 2007. AQUASTAT Database. Available online: www.fao.org/nr/water/aquastat/data/query/index.html

Food and Agricultural Organization of the United Nations (FAO). 2000. *New Dimensions in Water Security.* Rome: Land and Water Development Division.

Global Water Partnership (GWP). 2000. *Towards Water Security: A framework for action.* Stockholm: GWP.

Grey, D., and Connors, G. 2009. *The Water Security Imperative: We must and can do more.* Stockholm World Water Week. Available online: www.worldwaterweek.org/documents/WWW_PDF/Resources/2009_19wed/0903_Grey_Connors_The_Water_Security_Imperative_FINAL_PRESS.PDF

Grey, D., and Sadoff, C. 2007. Sink or Swim? Water Security for Growth and Development. *Water Policy* 9: 545–571.

International Commission on Large Dams (ICOLD). 2003. Paris: World Register of Large Dams.

International Water Management Institute (IWMI). 2009. *Flexible Water Storage Options: For adaptation to climate change.* Colombo, Sri Lanka: International Water Management Institute (IWMI). IWMI Water Policy Brief 31.

Lankford, B., Bakker, K., Zeitoun, M., and Conway, D. (Eds.) 2013. *Water Security: Principles, perspectives and practices.* London and New York: Routledge.

Liu, B., Mei, X., Li, Y., and Yang, Y. 2007. The connotation and extension of agricultural water resources security. *Agricultural Sciences in China* 6(1): 11–16.

Maplecroft 2011. *Water Security Risk.* Maplecroft, UK. Available online: http://maplecroft.com/about/news/water_security.html

Mitchell, T. D., Hulme, M., and New, M. 2002. Climate data for political areas. *Area* 34: 109–112.

Molden, D., Lautze, J., Shah, T., Bin, D., Giordano, M., and Sanford, L. 2010. Growing enough food without enough water: Second best solutions show the way. *International Journal of Water Resources Development* 26(2): 249–263.

Sadoff, C., Greiber, T., Smith, M., and Bergkamp, G. 2008. *Share: Managing water across boundaries.* Gland, Switzerland: IUCN.

Sinha, U. 2009. The why and what of water security. *Strategic Analysis* 33(4): 470.

Smakhtin, V., Revenga, C., and Döll, P. 2004. A pilot global assessment of environmental water requirements and scarcity. *Water International* 29(3): 307–317.

Swaminathan, M. 2001. Ecology and equity: Key determinants of sustainable water security. *Water Science and Technology* 43: 35–44.

Tarlok, A., and Wouters, P. 2009. Reframing the water security dialogue. *Journal of Water Law* 20: 53–60.

Taylor, R. 2009. Rethinking water scarcity: The role of storage. *Transactions of the American Geophysical Union* 90(28): 237–238.

United Nations University (UNU). 2013. *Water Security and the Global Water Agenda: A UN-Water analytical brief.* Ontario, Canada: United Nations University.

University of East Anglia. 2009. Water Security Research Centre. Available online: www.uea.ac.uk/watersecurity

Vorosmarty, C., McIntyre, P. B., Gessmer, M. O., Dudgeon, D., Prusevich, A., Green, P., Glidden, S., Bunn, S., Sullivan, C., Liermann, C., and Davies, P. 2010. Global threats to human water security and river biodiversity. *Nature* 467: 555–561.

WaterLink. 2010. *Nile Basin Initiative Deadlock.* July 6, 2010. Available online: www.waterlink-international.com/news/id1230-Nile_Basin_Initiative_Deadlock.html

World Health Organization (WHO), 2009. *World Health Statistics 2009.* Available online: www.who.int/whosis/whostat/2009/en/index.html

World Resources Institute (WRI). 2009. Earthtrends Environmental Information SEARCHABLE DATABASE. http://earthtrends.wri.org/#

Zeitoun, M. 2011. The global web of national water security. *Global Policy* 11: 1–11.

# 5 Water Productivity

*Jonathan Lautze, Xueliang Cai, and*
*Greenwell Matchaya*

## 5.1 Introduction

Improving water productivity (WP), especially in agriculture, is increasingly recognized as a central challenge in international development (Rockström et al., 2003; Comprehensive Assessment, 2007; USAID, 2009; CAADP, 2012; UNEP, 2012). The development research community (e.g., Rockström et al., 2003; CA, 2007) cast prominent attention on the potential for WP improvements to help unlock the economic potential of developing country farmers. FAO (2003) stated that improving agricultural water productivity is an important solution to addressing global water challenges. USAID's (2009) *Addressing Water Challenges in the Developing World: A Framework for Action* identified "improving water productivity" as one of its three core dimensions. The Comprehensive African Agricultural Development Program (CAADP) has given attention to the concept as a means to achieving its broader goals (CAADP, 2012). Finally, UNEP (2012) identified WP, as well as water use efficiency, as important indicators of water use in a green economy.

Despite widespread focus on improving WP, a growing body of literature (e.g., Bessembinder et al., 2005; Zoebl, 2006; van Halsema and Vincent, 2012; Ritzema, 2014) has emerged that identifies the concept's limitations. Bessembinder et al. (2005) highlight the need to normalize WP values and consider WP in conjunction with agricultural productivity. Zoebl (2006) suggests WP is not a useful concept in agricultural water management. Van Halsema and Vincent (2012) point to issues of scale when applying the WP indicator. Ritzema (2014) states that WP is not suited to contexts of multiple use systems with high water reuse and non-depleting water uses. While this literature marks valuable progress in circumscribing the roles in which WP should be utilized, it typically presupposes that WP holds value and seeks to delimit that value.

This chapter makes no such presupposition on WP's value, but rather seeks to compare the concept with related concepts of water efficiency[1] and agricultural productivity in order to identify whether the WP perspective offers benefits over and above those obtained through use of water efficiency and agricultural productivity. The chapter's main focus is on WP in agricultural water management. The chapter first reviews the origins of the concept as an outgrowth of irrigation efficiency (section 5.2). The chapter next explores the relationship between agricultural productivity and WP to identify redundancy and complementarity between the two concepts (section 5.3). Tools for improving WP are then sectorally disaggregated (section 5.4), as a means to revealing the particular value that the WP perspective adds (section 5.5). Inconsistencies and misconceptions associated with popular use and interpretation of the term are then explored (section 5.6). Finally, the chapter concludes (section 5.7) by suggesting that "improving WP" should not be treated as a central challenge in water management, but the WP indicator may hold value when employed together with other indicators.

## 5.2  From Irrigation Efficiency to Water Productivity

### Irrigation and Water Efficiency

A search for the origins of WP must start with focus on irrigation efficiency, out of which the WP concept sprang in the late 1990s. Irrigation efficiency is an indicator of the relationship between the amount of water required for a particular purpose and the amount of water applied for that purpose. It is measured as the proportion of water applied to a geographic unit such as a field, scheme, or system that is beneficially consumed (i.e., evapotranspired for an intended purpose). For example, if 10 cubic meters of water are applied in an irrigation system and 6 cubic meters are beneficially consumed by crops, irrigation efficiency is 0.6 or 60 percent (Seckler et al., 2003). Importantly, irrigation efficiency presumably includes field application efficiency, conveyance efficiency, and irrigation system efficiency.

Notably, the original indicator of irrigation efficiency was subsequently adapted to incorporate more nuance. The concepts of net efficiency (Jensen, 1977) and effective efficiency (Keller and Keller, 1995) were indeed created to attempt to account for potential downstream reuse of return flows of withdrawn water. In other words, these two indicators reflected an appreciation of the fact that a portion of the 4 in 10 cubic meters of water not depleted in the example above may be subsequently put to beneficial use. In effect, then, these indicators marked a shift from the fraction of *water beneficially consumed divided by water applied*, to *water beneficially consumed divided by total water consumed.*[2]

As basin-level approaches gained increasing prominence in the course of the 1990s, there was a growing desire to apply field-level efficiency concepts to the level of the river basin. Seckler et al. (2003) created the concept of basin efficiency as the rate of beneficial depletion of available water resources at a basin level. Computationally, this figure was calculated by dividing beneficial ET and beneficial flow to sinks by "available water supply."[3] Despite following intuitive logic in its development, the basin efficiency formula is at odds with more fundamental notions of irrigation efficiency, since it treats the concept as beneficially depleted water divided by all water in a basin *rather than* the beneficially depleted water divided by total water depleted. Anomalies associated with the basin efficiency formula were subsequently exposed (Perry, 2007) and, more broadly, anomalies associated with application of efficiency concepts at different scales received increasing spotlight (Molden and Sakthivadivel, 1999; Perry et al., 2009).

Importantly, to encompass the range of water-related efficiency indicators just described (irrigation efficiency, net efficiency, effective efficiency, basin efficiency), the chapter henceforth adopts the term "water efficiency."

## Water Efficiency and WP

In the context of the challenges of scaling up water efficiency concepts to a basin level, the WP concept was introduced. In broad terms, WP was deemed to be important due to perceptions of growing water scarcity and a looming "water crisis" (e.g., Molden, 1997; Cai and Rosegrant, 2003; Rijsberman, 2006). In response to existing and projected limitations on water supplies, it was postulated, more crops and other output needed to be produced with equal or less water. WP therefore measured the quantity of output per unit of water, which could be applied to assess performance toward the end of maximizing production derived from water use.

In addition to the broad logic for introducing WP, three sources (Cook et al., 2004; Turral et al., 2005; CA, 2007) provide more specific reasons for creating the WP indicator. Cook et al. (2004, p. 1) note that WP provides "a diagnostic tool to identify low or high water use efficiency in farming systems or sub-systems" in order to identify "opportunities for water redistribution within basins." Turral et al. (2005, p. 2) note that WP originated "broadly out of frustration with the ambiguity of concepts of irrigation efficiency." The Comprehensive Assessment for Water Management in Agriculture (2007, p. 283) echoes this sentiment, pointing out that the efficiency concept "provides only a partial and sometimes misleading view because it does not indicate the benefits produced."[4] These explanations allow one to interpolate that major drivers for creation of the WP concept were i) the desire to explicitly incorporate the benefits or output derived from water use, and ii) perceived limitations of efficiency indicators. It is nonetheless unclear if extensive thought was given to the additional problems that incorporating production derived from water use might engender, and whether limitations of efficiency indicators might be overcome by coupling efficiency results with those of agricultural productivity.

Whatever the case, definitions and measures for WP are abundant and the subject of frequent discussion (Molden, 1997; Kijne et al., 2003; FAO, 2003; Perry, 2007). In agriculture, WP has been described as a robust measure of the ability to convert water into food (Kijne et al., 2003). It has been expressed more rigorously as the total agricultural benefit per unit of water used, and more simply as crop-per-drop (Cook et al., 2004). In rainfed agriculture, WP has been described as "the efficiency of water use at the production system or farm level" (Rockström et al., 2003, p. 146). The concept has also expanded beyond agriculture to include multiple sectors, for example, according to the following definition: "the ratio of net benefits from crop, forestry, fishery, livestock, and agriculture systems to the amount of water required to produce those benefits" (Molden et al., 2010, p. 528).

In irrigated agriculture, WP is typically measured as either i) agriculture yield divided by water depletion, or ii) agriculture yield divided by water withdrawal (Kijne et al., 2003). There is increasing use of the former rather than the latter measure. In rainfed agriculture, WP is alternatively measured as i) agricultural production divided by rainfall, ii) agricultural production divided by effective rainfall, or iii) agricultural production divided by harvested water (Rockström et al., 2003).[5] Of these three measures, the one growing in frequency of use (Cai and Rosegrant, 2003; Igbadun et al., 2005) may be agricultural production divided by effective rainfall.

Direct comparison of two common indicators of WP with those of water efficiency (Table 5.1) reinforces the broader nature of WP. While the denominators of the common measures are identical, the numerators shift from "water beneficially depleted" to "agricultural production." Indeed, whereas water efficiency is focused on variables more directly tied to water use, the WP indicator includes the quite broad variable of agricultural production and the specific variable of water use. Notably, there is increasing use of the indicators shown in the bottom row of Table 5.1—i.e., there is increasing use of denominators focused on water depleted rather than water withdrawn. A likely driver for this shift from water withdrawal to depletion is the desire for an indicator that can be applied equally in irrigated and rainfed agriculture.[6]

Notable work has been conducted on livestock and fish WP, as well as intersectoral WP. Amede et al. (2009), for example, highlighted the benefits of incorporating livestock production into conventional measures of WP; Peden et al. (2009) focus mainly on a framework and options for raising livestock WP. Analogously, Dugan et al. (2006) highlight the fish production benefits derived from water use, yet fail to calculate any values for fish WP. Computation of fish WP would appear particularly incongruous with other

*Table 5.1* Water efficiency and water productivity in agricultural water management

|  | *Water efficiency* | *Water productivity* |
| --- | --- | --- |
| Common measure 1 | Water beneficially depleted/ water withdrawn | Agricultural production/ water withdrawn |
| Common measure 2 | Water beneficially depleted/ total water depleted | Agricultural production/ total water depleted |

approaches to WP since fish production does not require water to be withdrawn or consumed. Finally, Prasad et al. (2006) compared WP across the agriculture, industry, mining, and water supply service sectors in the Olifants basin of South Africa; not surprisingly, WP is highest in industry and lowest in agriculture.

## 5.3   Agricultural Productivity and Water Productivity: Redundant or Complementary?

Inadvertently or not, the broadening of water efficiency to WP engendered some level of duplication with the pre-existing concept of agricultural productivity. Such duplication or redundancy, in turn, triggers questions about the necessity of an independent WP concept; for example, were the benefits of WP already achieved through use of agricultural productivity? While it may be that scrutiny of the two concepts reveals a complementary niche for each, it may also be that comparison of agricultural productivity and WP renders the WP concept moot. This section examines meanings and measures of agricultural productivity relative to WP, in order to determine whether the two concepts are complementary, redundant, or somewhere in between.

### Agricultural Productivity

Agricultural productivity is defined as the ratio of agricultural product to agricultural inputs (Zepeda, 2001). The origins of the agricultural productivity concept can be traced to the 1950s (or earlier) when limitations on increasing agricultural production through expansion of land use spurred interest in raising productivity on existing lands (Wiebe et al., 2001). Initially, the two most common ways to measure agricultural productivity were i) agricultural production per unit of land, and ii) agricultural production per unit of labor. Increasingly, however, the strength of these indicators has been questioned on the grounds that they capture only a small portion of the factors explaining agricultural production outcomes.

To account for variation in the explanatory power of productivity indicators, measures of agricultural productivity are presently divided according to whether they capture i) only one input that explains agricultural production, ii) all inputs that explain production. Evaluating just one factor of production with agricultural output—i.e., the former group— is called partial factor productivity. To calculate partial factor productivity, total agricultural production is typically divided by just one factor of production. For example, the weight or value of crops produced can be summed and divided by the area on which they were cultivated to determine agricultural productivity relative to the factor of land. Comparing all factors of production with agricultural output—i.e., the latter group—is called total factor productivity (TFP). To calculate TFP, the value of inputs to agricultural production are monetized and aggregated, and the value of total agricultural output is monetized and aggregated. The total value of agricultural output is then divided by aggregate value of inputs. It should be noted that whereas all inputs are theoretically considered in agricultural TFP calculations, in practice difficulties associated with monetization of water limit its inclusion.

While there is increasing use of TFP as a definitive measure of agricultural productivity, one measure of partial factor productivity that remains widely used is production per unit of land. Land productivity—also known as *yield* on agriculture lands—is used by policy makers to assess impacts of new production practices, and to monitor change in agricultural production to meet national food security needs (Wiebe et al., 2001). Use of the land productivity concept is nonetheless often qualified by an explanation that determination of agricultural productivity based on only one factor may be misleading, as it fails to provide definitive indication of relative contribution of the multiple factors that explain levels of production. An examination of variation in levels of agriculture yields per hectare, for example, would fail to reveal whether that variation is explained by fertilizer, crop selection, mechanization, water use, or other factors.

### Agricultural Productivity and Water Productivity

Agricultural productivity, in a broad sense, is a holistic term that includes all the factors of production for a given enterprise during a particular season under a specified environment. Since water comprises one factor of agricultural production, WP can be considered a measure of partial factor productivity. Notably, variations in water management (deficit-irrigation, full irrigation, micro-irrigation, level of waterlogging, etc.) may be reflected in the values of agricultural productivity. However, variation in the quantity of water used is rarely captured in TFP calculations because water is often not priced.

Water's exclusion from TFP calculations suggests that, from a practical perspective, WP and TFP are complementary. Whereas TFP would seem to capture the quantities of all inputs to agricultural production except water, WP incorporates precisely the part of the picture that TFP omits. Further, just as there is utility in capturing the aggregate value of all non-water inputs to agricultural production in TFP, there would appear value in understanding how much water is depleted to achieve agricultural output.

It is worth noting that combining agricultural and water productivity results has been advocated for, and practically applied. Bessembinder et al. (2005), for example, called for joint use of agricultural productivity and WP as a means to identifying the optimum blend of production per hectare and production per $m^3$ of water. Cai and Sharma (2010) mapped both WP and yield in the Indo-Gangetic basin to narrow down options for making agricultural and water productivity improvements. There would therefore appear potential to couple these two indicators so as to increase the value of the information resulting from their use.

At least part of the reason for coupling WP and agricultural productivity results in these examples is to minimize potential for isolated use of WP to generate misleading results. Similar to other measures of partial factor productivity, the fundamental limitation of the WP indicator would appear to be the number of variables that it omits. As noted, there are numerous factors (e.g., seeds, fertilizer, labor) that affect agricultural production, and application of WP in isolation fails to reveal the relative contribution of the various factors—including water—to WP results. As such, while the WP concept would appear to hold value as a complement to TFP, WP's value when used in isolation is likely to mask numerous important variables. Factors that explain variation in WP outcomes may indeed have no relation to water.[7]

## 5.4 Water Productivity: Duller Messages and Murkier Guidance?

At a practical level, the most important test of WP's added value is whether use of the concept enables identification of specific solutions, interventions, or policy options that could not be determined through use of pre-existing concepts of water efficiency and agricultural productivity. To justify creation of a new concept, it would indeed help to know that application of that concept sheds light on a possible response to a development challenge that could not otherwise be identified. This section therefore considers whether interventions that raise WP are distinct from those that improve water efficiency and agricultural productivity.

To identify measures that improve WP, various sources were reviewed (Kijne, 2003; Rockström et al., 2003; Cook et al., 2004; Zwart and Bastiaanssen, 2007; CA, 2007; Perry et al., 2009; van Halsema and Vincent, 2012). There appeared notable disparity in lists of measures, strategies, or options for improving WP. Molden et al. (2003, p. 11), for example, lists eleven strategies to improve WP, and only one of these strategies gives attention to non-water inputs. Van Halsema and Vincent (2012, p. 12), by contrast, list three factors that determine WP at a farm level: "(i) crop type and crop genetics; (ii) nutrient deficiencies in the crop growth cycle (e.g. nutrient deficiencies equate to lower WP), and (iii) to a lesser extent, irrigation application and cultivation techniques that affect evaporation." To overcome these disparities, a long list of potential measures was created based on a review of relevant sources. Measures were then consolidated when there appeared substantial overlap, and grouped into five categories: crop-related, chemical-related, economy-related, soil-related, water-related. The resulting list of grouped measures (Table 5.2), while likely not exhaustive, is believed to reflect most of the major interventions for raising WP.

*Table 5.2* Measures to improve water productivity

| Grouping of measures | Measure that improves WP |
| --- | --- |
| Crop-related | – Crop type<br>– Crop genetics<br>– Mechanization |
| Chemical-related | – Herbicide<br>– Insecticide |
| Economy-related | – Labor increase and strengthening (e.g., extension)<br>– Market access<br>– Post-harvest food storage<br>– Water pricing |
| Soil-related | – Fertilizer<br>– Compost<br>– Reduced tillage<br>– Crop rotation |
| Water-related | – Reducing un-beneficial water depletion (i.e., water losses)<br>– Supplemental irrigation<br>– Deficit irrigation<br>– Drip/sprinkler irrigation (micro-irrigation)<br>– Maintenance and upgrade of water infrastructure<br>– Irrigation scheduling<br>– Water harvesting<br>– Water reuse |

While the list of measures (Table 5.2) may not be exhaustive, it none-theless likely captures most major tools for raising WP. The list generates at least two broad findings that are unlikely to be overturned even if additional measures are identified. A first broad finding is that the majority of measures for raising WP are not directly related to water. In other words, there are more options for raising WP through interventions outside the water sector. While this reality may ultimately highlight the merits of utilizing integrated approaches that draw on measures from multiple sectors, it is nonetheless worth noting that the proportion of water sector interventions that produce improvements in a water sector indicator may be smaller than expected.

A second broad finding is that no measure *uniquely* improves WP. In other words, scrutinizing the list of measures that improve WP suggests that any measure employed to raise WP can also be classified as either: i) a measure that improves agricultural productivity, ii) a measure that improves water efficiency. Interventions related to crop genetics, fertilizer or compost typically raise production and hence improve agricultural productivity, for example, and interventions focused on deficit irrigation, irrigation scheduling, and water reuse improve water efficiency. As such, it would seem that this list of WP measures could alternatively be labeled as a list of measures that improve water efficiency and agricultural productivity.

From a policy perspective, it would appear that sharper messages and more targeted guidance are achieved through use of water efficiency or agricultural productivity rather than WP. Indeed, whereas use of pre-existing indicators of water efficiency and agricultural productivity directly reveal which area needs improvement, use of the WP concept only guides a user to identifying that improvements in *either* water efficiency *or* agricultural productivity are needed. As a result, as the WP indicator fails to reveal whether to implement measures to improve agricultural productivity or water efficiency, no directly actionable recommendations are evident. A review of WP literature (e.g., Zwart and Bastiaanssen, 2004; Prasad et al., 2006; Zwart and Bastiaanssen, 2007; Cai et al., 2011) indeed confirms that use of WP results in isolation typically fail to enable tailored recommendations.[8]

## 5.5 The Bottom Line: Is There Value in Water Productivity?

Before reaching an ultimate verdict on the technical value added by WP, at least three softer benefits of the WP concept merit mention. First, WP has likely served a bridge between disciplines and sectors, leading to more interaction between agriculture and water experts and likely fostering more integrated approaches. Second, the creation of the WP indicator likely helped raise the profile of water and water management in discussions of agriculture and food security, possibly owing to a certain allure of the WP concept. Third, related to the last point, WP has likely helped increase appreciation for the essential role of water in agricultural production.[9] Insofar as use of WP helps attract more attention to this fact, the concept can be considered to hold value.

It must nonetheless be acknowledged that agricultural production is affected by many variables in addition to those found in the simple water productivity formula. While all indicators are to some extent a simplification of reality, the list of rather major variables that affect agricultural output that are not contained in the WP indicator—e.g., quality of seeds, quantity of fertilizer used, application of insecticide—is quite lengthy. Use of the water productivity indicator would indeed be tantamount to running a single variable regression when there may in fact be five or more major variables affecting an outcome.

Ultimately, determination of WP's added value may be revealed by focusing on the roles in which the concept was envisioned to be applied. Primary roles to which WP was envisioned to be applied can be distilled to just two:

1. assessing performance toward the end of maximizing production from water, which could be used to identify opportunities for raising production derived from water in particular locations (i.e., areas where WP is low);
2. assessing performance toward the end of maximizing production from water, which could be used as a guide to decision making about allocating water between sectors, schemes, or plots (i.e., re-allocating from low value uses to high value uses).

The fundamental constraint on use of WP in the first role is that the indicator fails to provide any specific guidance on how to improve conditions. That is, identification that an area possesses low WP would indeed indicate that productivity can be improved in that area. However, the course of action that should be followed to raise WP would not be clear as use of the WP indicator fails to reveal whether to apply water management measures or agricultural tools. While one constructive option here may be to apply the WP indicator together with water efficiency and agricultural productivity indicators in order to identify specific actions that should be taken, this then begs the question as to whether use of the WP indicator adds additional value to joint use of water efficiency and agricultural productivity.

Related to use of WP in the second role, at least two reservations should be flagged. First, variation in WP across sectors would likely carry few surprises given the relative GDP of different sectors. Second, use of WP as guide to decision making about water allocation between one agricultural scheme and another fails to consider the degree to which such WP levels are inherent v. the result of any of a myriad of manageable factors affecting production. If in fact low WP is due to physical constraints, the WP indicator may serve a useful guide to redirecting the resource to where it is likely to be more productive. However, if low WP is due to poor decision making associated with manageable parameters, allocating more water to areas of higher WP (or directing water to higher valued uses) would seem to dismiss the potential to raise WP where it is low.

The bottom line is that in the first role identified above, it appears that the WP perspective *does not* add value over and above joint use of water efficiency and agricultural productivity. Opportunities to improve WP in a particular location can indeed be identified with a greater degree of specificity through joint use of water efficiency and agricultural productivity indicators. In the second role identified, it appears that the WP perspective *does* add value over and above the two other concepts. Joint use of water efficiency and agricultural productivity do not inform decision making to the same degree as WP because other indicators fail to enable comparison of the production benefits associated with use of water in one location v. another. WP can thus serve as a decision-making guide to water allocation.

Substantial qualification should nonetheless be attached to WP's use in the second role, as application of the indicator fails to reveal whether WP results are due to inherent constraints (e.g., a region's climate) or manageable parameters. That said, combining use of WP with other indicators would help to reduce the spectrum of potential contributing factors and increase our ability to identify specific changes that would positively affect conditions. The bottom line is that WP holds value when employed as a qualified guide to gauging production derived from water use in different locations and sectors, which one should apply in conjunction with other indicators.

## 5.6  Discussion: Rhetorical Value Meets Practical Limitations

This chapter has compared definitions and indicators of water efficiency, WP, and agricultural productivity in order to identify the value and benefits of WP over and above those provided by the other indicators. The chapter's main finding is that WP indeed holds value as a decision-making guide for allocating water between areas and sectors. The chapter nonetheless also found that WP should be applied with other indicators, and use of WP as a standalone indicator holds potential to mislead. Further, the chapter found that WP does not produce added value when applied in a particular location such as a scheme or farm; pre-existing indicators of water efficiency and agricultural productivity may in fact prove more useful at this scale.

These findings are consistent with some literature, and inconsistent with other literature. Consistent with this chapter's finding that WP should be utilized in conjunction with other indicators, for example, Bessembinder et al. (2005) call for use of WP in conjunction with agricultural productivity. Further, consistent with this chapter's finding that WP is useful at larger scales in which WP at different locations can be compared, van Halsema and Vincent (2012) suggest irrigation efficiency may be a more useful concept at a scheme level and WP at a basin scale. While this chapter determined the utility of WP to be confined to one role and best used in that role in conjunction with other indicators, this indicates that the chapter found WP to in fact hold value. As such, the chapter's findings are at odds with those of Zoebl (2006), who suggested the concept holds no value.

A broader issue relates to the implications of our findings on the role for WP. As just noted, the chapter determined WP to be an indicator suited to one role, in which the concept should not be applied as a standalone measure. This tightly confined use of WP likely conflicts with some other notable documents (e.g., Giordano et al., 2006; Rijsberman, 2006), which declare WP as a "paradigm" or "framework" for water resources management. The delimited role identified in the chapter is also at odds with policy and development documents (e.g., FAO, 2003; USAID, 2009) that treat improving WP as a central challenge in water resources management.

The delimited role nonetheless appears consistent with original use of the indicator. It should be recalled that WP originated as a concept (Molden,

1997) specific to irrigation, spurred largely by limitations of irrigation efficiency. In this original role, WP was intended to be applied as one indicator with other indicators as part of a broader water accounting framework. Yet, as the concept's use increased, it appeared to assume a more "paradigmatic" status, through which it was treated as a central goal for water resources management and exported to areas beyond irrigation (e.g., rainfed agriculture, livestock, fish).

Ultimately, the gap between WP's rhetorical promise and practical limitations may in fact be responsible for the concept's misuse, as WP's catchiness may have fostered a desire to inject the theoretical term into practical roles and discussions that went beyond the term's capacities and resulted in increased confusion. The desire to insert WP into pre-existing discussions is apparent, for example, from exploring the extent to which the concepts of water efficiency and WP have been used interchangeably in many journal articles (see Perry, 2007, p. 375 for a review), and an examination of the extent to which the language of efficiency is incorporated into descriptions of WP (e.g., Rockström et al., 2003; Cook et al., 2004—see section 5.2). While these examples may reflect a desire to repackage old discussions of water efficiency into the new framework of WP, dangers arise because differences between water efficiency and WP are more than semantic. It is indeed possible for water efficiency to decrease and WP to increase, or vice versa. For example, water could be used more wastefully leading to more un-beneficial ET and lower efficiency, yet WP could still improve so long as efficiency decreases are more than offset through agricultural production increases.[10] Conversely, efficiency could improve through reductions in water losses, yet WP could decrease as a result of lower agricultural production levels resulting from, for example, the diminishing returns of fertilizer use.

There may have also been a desire to subsume old water management techniques under the heading of the new WP framework, yet it's worth noting that some of these measures do not *necessarily* improve WP. For example, use of measures such as supplemental irrigation and ridge tillage have been associated with WP improvements, and to be clear, often do lead to WP improvements. Nonetheless, adding more water to produce more crops—i.e., more drops, more crops—does not necessarily mean that the ratio of water to agriculture production improves. As such, one cannot definitively say that implementation of either of these activities raises WP. On the contrary, one can say that implementing these activities generally raises agricultural production and productivity.

### 5.7  Conclusion: From Central Paradigm back to Routine Indicator

This chapter has applied critical analysis to the meaning and value added of WP. The chapter's findings suggest WP should not be treated as a central paradigm but rather should be treated as one indicator that is best employed as one among many; for example, as evidenced in Karimi et al.'s (2013) Water Accounting Plus framework. The chapter has nonetheless postulated that the WP concept may carry rhetorical or strategic value to the water community. That is, as food security and agriculture rose to prominence in the international development community, with important implications for allocation of funding, a water paradigm that carried obvious linkages to food and agriculture may have served a strategic purpose. Finally, the chapter has questioned whether the profile-raising value of WP may have engendered increased use of the concept in practical roles to which it was not suited, leading to confusion.

One issue yet to receive focus is the short- v. long-term impact of WP's rhetorical use. Indeed, while promotion of WP may have helped to attract attention and funding to the water sector in the short term, the extent to which WP's prominence in development circles will continue to attract such attention and funding in the longer term remains to be determined. Clearly, an indicator such as WP that reflects the performance of multiple sectors holds potential to capture the effects of integrated approaches so widely promoted—which is a positive. It nonetheless appears that most measures to improve WP lie outside the water sector. As such, it may be that WP raises the profile of the water sector, only to direct development practitioners to a suite of interventions of which the majority are outside the water sector. Such an outcome would appear inconsistent with one of the concept's purposes of promoting the water sector in development discussions.

Ultimately, the role for water management is to *enable* rather than *determine* agricultural production. As such, the ideal water sector paradigm might seek to capture how effectively provision of water matches crop water requirements, as a means to *enabling* agricultural production and productivity. Notably, however, wide variation in agricultural productivity is likely to be evidenced in areas of equally effective provision of water—owing to the multitude of non-water factors that contribute to agricultural production. While production outcomes associated with allocation of water may be used as indicative input to decision making in planning models, there would appear a simultaneous need to separate indicators in the water sector from those capturing agricultural production outcomes, and a need to independently measure the effectiveness of water resources management.

Before concluding, it is worth noting that there is no perfect indicator here, and that indicators of both efficiency and productivity fail to capture a rather major practical benchmark of water management effectiveness: the degree to which water is provided where and when it is needed. Further, achieving high levels of water efficiency and productivity may be a somewhat academic pursuit in regions where water is abundant. A useful alternative to efficiency and productivity might therefore focus on measuring the degree to which water supply matches water demand at a frequency that is relevant to the requirements of productive water uses. Ultimately, ensuring that provision of water is matched to the uses of that water may be more critical to food security, economic growth, and other objectives with which water efficiency and WP are often associated.

In conclusion, this chapter has determined that use of the WP indicator may be best confined to a water allocation guidance tool that is employed in conjunction with a broader set of indicators. In addition, promoting a target of *improving* WP as a paradigm central to water management is somewhat dubious, given the fact that the WP indicator reflects the performance of so many variables that are extraneous to water management. These conclusions strongly suggest that the WP concept should be demoted from the status of "paradigm" or "framework," back to a status of simple "indicator."

## Notes

1. Please note that use of the word water efficiency is formally defined below to encompass water-related efficiency concepts such as: irrigation efficiency, net efficiency, effective efficiency, basin efficiency.
2. Please note that "consumption" and "depletion" are used interchangeably in this chapter.
3. Basin Efficiency = (Eb+Sb)/AWS. That is, basin efficiency is equal to beneficial evaporation plus beneficial flow to sinks divided by annual water supply.
4. Identical text can be found in Molden et al. (2010).
5. It is worth noting here that WP's increasing application to areas of rainfed agriculture may have led the concept to assimilate elements of water use efficiency as used in agronomy. Agronomic water use efficiency—measured as kg of biomass produced per $m^3$ of water used—is focused on the ability of a plant to achieve maximum yield with minimal water used. While it is notable that water use efficiency incorporates the benefits of water use similar to WP, there nonetheless remain two fundamental differences between water use efficiency in agronomy and water sector concepts of WP and water efficiency. First, water use in agronomic calculations is generally in the sense of water used by a plant (i.e., transpired) rather than total water used (i.e., depleted—lost to a system). Second, the scales at which the agronomic and water-sector concepts are intended to be applied vary substantially: water use efficiency at a plant level, WP and water efficiency at much broader levels.
6. An additional, more practical reason for the increased use of water depletion may be the abundance of remote sensing and GIS tools available to estimate the total water depletion at a basin scale.
7. While this last point may seem mundane to some, it is worth pointing out that certain sources seem to focus heavily on water sector options for improving WP (e.g., Molden et al. 2003, page 11—discussed below).
8. While studies that were reviewed do provide recommendations, such recommendations are typically *cookie-cutter* in nature, because variations in WP scores rarely provide specific guidance on how to improve productivity in a particular location.
9. Indeed, while those in the water sector may treat water as an obvious part of agriculture development and food security efforts, there may be a tendency in other sectors to take water for granted and place more emphasis on other factors that contribute to agriculture production.
10. For example, through use of improved seeds and fertilizer.

# References

Amede, T., Geheb, K., and Douthwaite, B. 2009. Enabling the uptake of livestock–water productivity interventions in the crop–livestock systems of sub-Saharan Africa. *The Rangeland Journal* 31: 223–230.

Bessembinder, J. J. E., Leffelaar, P. A., Dhindwal, A. S., Ponsioen, T. C. 2005. Which crop and which drop, and the scope for improvement of water productivity. *Agricultural Water Management* 73(2): 113–130.

CAADP, 2012. *Improving Water Productivity and Efficiency Fact Sheet.* Climate-Smart Agricultural Workshop. Nairobi, Kenya. Available online: www.caadp.net/pdf/2a.%20Improving%20Water%20Productivity%20and%20Efficiency.pdf

Cai, X., and Rosegrant, M. 2003. World water productivity: Current situation and future options. Chapter 10 in: *Water Productivity in Agriculture: Limits and Opportunities for Improvement.* Wallingford, UK, and Colombo: CABI Publishing and International Water Management Institute.

Cai, X., and Sharma, B. 2010. Integrating remote sensing, census, and weather data for an assessment of rice yield, water consumption, and water productivity in the Indo-Gangetic river basin. *Agricultural Water Management* 97: 309–316.

Cai, X., Molden, D., Mainuddin, M., Sharma, B., Ahmad, M., and Karimi, P. 2011. Producing food with less water in a changing world: Assessment of water productivity in 10 major river basins. *Water International* 36(1): 42–62.

Comprehensive Assessment of Water Management in Agriculture. 2007. *Water for Food, Water for Life: A comprehensive assessment of water management in agriculture.* London: Earthscan, and Colombo: International Water Management Institute.

Cook, S., Gichuki, F., and Turral, H. 2004. *Water Productivity: Estimation at a plot, farm and basic scale.* Basin Focal Project Working Paper 2. Colombo: CGIAR Challenge Program on Water and Food. Available online: http://ageconsearch.umn.edu/bitstream/91959/2/H039742.pdf

Dugan, P., Dey, M., and Sugunan, V. 2006. Fisheries and water productivity in tropical river basins: Enhancing food security and livelihoods by managing water for fish. *Agricultural Water Management* 80(1–3): 262–275.

FAO, 2003. Unlocking the water potential of agriculture: Why agricultural water productivity is important for the global challenge. Chapter 3 in: *Unlocking the Water Potential of Agriculture.* Food and Agricultural Organization of the United Nations. Available online: www.fao.org/docrep/006/y4525e/y4525e00.htm#Contents

Giordano, M., Rijsberman, F., and Saleth, R. 2006. "More Crop Per Drop": Revisiting a Research Paradigm. London: IWA Publishing.

Igbadun, H., Mahoo, H., Tarimo, A., and Salim, B. 2005. *Trends of Productivity of Water in Rain-fed Agriculture: Historical perspective.* Colombo: International Water Management Institute.

Jensen, 1977. Water conservation and irrigation systems: Climate. In: *Technical Seminar Proceedings*, Columbia, Missouri, pp. 208–230.

Karimi, P., Bastiaanssen, W. G. M., and Molden, D. (2013). Water Accounting Plus (WA+)—a water accounting procedure for complex river basins based on satellite measurements, *Hydrology and Earth System Sciences* 17: 2459–2472.

Keller, A., and Keller, J. 1995. Effective efficiency: A water use efficiency concept for allocating freshwater resources. Brief for the Center for Economic Policy Studies. Arlington, VA: Winrock International.

Kijne, J., Barker, R., and Molden, D. (Eds.). 2003. *Water Productivity in Agriculture: Limits and opportunities for improvement*. Wallingford, UK, and Colombo: CABI Publishing and International Water Management Institute.

Molden, D. 1997. *Accounting for Water Use and Productivity*. SWIM Paper 1. Colombo, Sri Lanka: International Irrigation Management Institute.

Molden, D. and Sakthivadivel, R. (1999) Water accounting to assess uses and productivity of water. *Water Resources Development* 155(1–2): 55–71.

Molden, D., Murray-Rust, D., Sakthivadivel, R., and Makin, I. (2003) A water productivity framework for understanding and action. Chapter 1 in: J. Kijne, R. Barker, and D. Molden (Eds.) *Water Productivity in Agriculture: Limits and opportunities for improvement*. Wallingford, UK, and Colombo: CABI Publishing and International Water Management Institute.

Molden, D., Oweis, T., Steduto, P., Bindraban, P., Hanjra, M., and Kijine, J. 2010. Improving agricultural water productivity: Between optimism and caution. *Agricultural Water Management* 97: 528–535.

Peden, D., Taddesse, G., Halieslassie, A. 2009. Livestock water productivity: Implications for sub-Saharan Africa. *The Rangeland Journal* 31: 187–193.

Perry, C. 2007. Efficient irrigation: Inefficient communication, flawed recommendations. *Irrigation and Drainage* 56: 367–78. Wiley Interscience.

Perry, C., Steduto, P., Allen, R., and Burt, C. 2009. Increasing productivity in irrigated agriculture: Agronomic constraints and hydrological realities. *Agricultural Water Management* 98: 1517–1524.

Prasad, K., van Koppen, B., and Stzepeck, K., 2006. Equity and productivity assessments in the Olifants river basin, South Africa. *Natural Resources Forum* 30(1): 63–75.

Rijsberman, F. 2006. "More crop per drop": Realigning a research paradigm. Chapter 2 in: M. Giordano, F. Rijberman, and R. Saleth (Eds.) *"More Crop Per Drop": Revisiting a research paradigm*. London: IWA Publishing.

Ritzema, R. 2014 in press. Aqueous valuation: An alternative productivity indicator for water. *Hydrology*.

Rockström, J., Barron, J., and Fox, P. 2003. Water productivity in rain-fed agriculture: Challenges and opportunities for small-holder farming in drought-prone tropical agroecoystems. In: J. Kijne, R. Barker, and D. Molden (Eds.) *Water Productivity in Agriculture: Limits and opportunities for improvement*. Wallingford, UK, and Colombo, Sri Lanka: CABI Publishing and International Water Management Institute, pp. 145–162.

Seckler, D., Molden, D., and Sakthivadivel, R. 2003. The concept of efficiency in water resources management and policy. In J. W. Kijne, R. Barker, and D. Molden (Eds.). *Water Productivity in Agriculture: Limits and opportunities for improvement*. Wallingford, UK, and Colombo: CABI Publishing and International Water Management Institute, pp. 37–51.

Turral, H., Cook, S., and Gichuki, F. 2005. Water productivity assessment: Measuring and mapping methodologies. *Basin Focal Project Working Paper No. 2*. Colombo: CGIAR Challenge Program on Water and Food. Available online: http://r4d.dfid.gov.uk/PDF/Outputs/WaterfoodCP/AgriculturalWaterProductivity BFPwp02Draft03.pdf

UNEP, 2012. *Measuring Water Use in a Green Economy*. A Report of the Working Group on Water Efficiency to the International Resource Panel. McGlade, J., Werner, B., Young, M., Matlock, M., Jefferies, D., Sonnemann, G., Aldaya, M.,

Pfister, S., Berger, M., Farell, C., Hyde, K., Wackernagel, M., Hoekstra, A., Mathews, R., Liu, J., Ercin, E., Weber, J. L., Alfieri, A., Martinez-Lagunes, R., Edens, B., Schulte, P., von Wirén-Lehr, S., and Gee, D. Available online: www.unep.org/resourcepanel/Portals/24102/Measuring_Water.pdf

USAID, 2009. *Addressing Water Challenges in the Developing World: A framework for action.* Washington, DC: Bureau for Economic Growth Agriculture and Trade.

van Halsema, G. E., and Vincent, L. 2012. Efficiency and productivity terms for water management: A matter of contextual relativism versus general absolutism. *Agricultural Water Management*, 108: 9–15.

Wiebe, K., Soule, M., and Schimmelpfennig, D. 2001. Agricultural productivity for sustainable food security in sub-Saharan Africa. Chapter 4 in: L. Zepeda (Ed.) *Agricultural Investment and Productivity in Developing Countries.* Rome, FAO.

Zepeda, L. 2001. Agricultural investment, production capacity and productivity. Chapter 1 in: L. Zepeda (Ed.) *Agricultural Investment and Productivity in Developing Countries.* Rome: FAO.

Zoebl, D. 2006. Is water productivity a useful concept in agricultural water management? *Agricultural Water Management* 84: 265–273.

Zwart, S., and Bastiaanssen, W. 2004. Review of crop water productivity values for wheat, rice, cotton, and maize. *Agricultural Water Management* 69: 115–133.

Zwart, S., and Bastiaanssen, 2007. SEBAL for detecting spatial variation of water productivity and scope for improvement in eight irrigated wheat systems. *Agricultural Water Management* 89: 287–296.

# 6  Virtual Water and Water Footprints

*Dennis Wichelns*

## 6.1  Introduction

The term "virtual water" began appearing in the water resources literature in the mid 1990s. Professor Tony Allan chose the term to describe the water used to produce crops traded in international markets (Allan, 1996, 2002). The notion of "water footprints" appeared a bit later, put forth originally in conjunction with the discussion of the amounts of virtual water "flowing" between countries as they trade goods and services (Hoekstra and Hung, 2002; Chapagain et al., 2006a, 2006b). In recent years, many authors have calculated the volumes of virtual water and water footprints within countries (Ma et al., 2006; Guan and Hubacek, 2007), while others have calculated both "internal" and "external" water footprints, suggesting that these measures distinguish between how much virtual water is used within a country from the volume involved in international trade (Hoekstra and Chapagain, 2007a, 2007b). Green, blue, and grey water footprints also have been described (Van Oel et al., 2009; Fader et al., 2011). Green water footprints are intended to represent the soil moisture from rainfall that is used for crop production, while blue water footprints describe irrigation with surface water or groundwater (Mekonnen and Hoekstra, 2010). The grey water footprint is offered as an estimate of the volume of water required to dilute pollutants to meet the prevailing ambient water quality standards (Hoekstra, 2013: 11).

The original promoters of virtual water and water footprints might have intended to use the terms to increase awareness and bring helpful attention to important water resource issues. Indeed, the terms likely have succeeded in promoting greater awareness of the role of water in the production of many goods and services. In water scarce areas, such an outcome is certainly desirable. Yet, even in water scarce areas, water is just one of many inputs. In many settings, producers, consumers, and public officials must consider issues that extend beyond water, when crafting public policies or when determining optimal production and consumption strategies.

The goal of this chapter is to determine whether the notions of virtual water and water footprints are policy relevant. The popularity of the notions is evident in the large number of articles published in the popular press and in professional journals. In addition, several large companies, including Coca-Cola, Pepsico, SABMiller, and Unilever, have begun calculating the water footprints of selected products (SABMiller and WWF-UK, 2009; Hoekstra, 2013). Discussions of water footprints are appearing also in national legislatures and policy documents.

In April 2012, the Dutch House of Representatives considered a motion requesting

> that the government, in its economic policy, aim for Dutch companies to present their water footprint and to reduce this footprint in those areas that are affected by water scarcity, for example, by actively addressing companies that receive support through export guarantees or innovation subsidies to reduce their water footprints, and to request that these companies calculate their water footprints and include this information in their sustainability reports.
>
> (Witmer and Cleij, 2012, p. 40)

Furthermore, the motion requested "that the government, on an EU level, during the reform of the Common Agricultural Policy, will aim to reduce the subsidising of water-intensive agriculture in areas of water scarcity, such as cotton production in Mediterranean countries" (Witmer and Cleij, 2012, p. 40). In Spain, the Ministry of the Environment requires the use of water footprints in developing river basin plans, in efforts to achieve compliance with the European Union's Water Framework Directive (Aldaya et al., 2010).

In April 2013, the Ministry of Water Resources convened a special seminar on water footprints in New Delhi as part of India Water Week. Several statements regarding water footprints appear in India's new National Water Policy (Government of India, 2012). The statements, which appear in the section pertaining to demand management and water use efficiency, are the following:

1. A system to evolve benchmarks for water uses for different purposes; i.e., water footprints, and water auditing should be developed to promote and incentivize efficient use of water.
2. The project appraisal and environment impact assessment for water uses, particularly for industrial projects, should, *inter-alia*, include the analysis of the water footprints.

Promoted originally as attractive indicators of the amount of water required to produce a good or service, it appears water footprints are now being considered and adopted as policy tools in national legislation. These examples from India, Spain, and the Netherlands might be the first of many cases in which governments consider requiring firms to measure and reduce their water footprints. It seems fair to ask whether such an approach is supported by an underlying conceptual framework or a persuasive empirical record.

## 6.2  Literature Review

We examine three issues in a review of literature pertaining to virtual water and water footprints: 1) the pertinence of virtual water in discussions regarding international trade, 2) the notion of saving or losing water by organizing production patterns according to water footprints, and 3) the importance of considering the livelihood impacts of water resource policies, particularly in developing countries.

### Regarding International Trade

Several authors of the literature on virtual water suggest or imply that water-short countries should import water-intensive products from countries with larger water endowments, and that water-abundant countries should focus on producing and exporting water-intensive goods (Yang and Zehnder, 2002; Hoekstra and Hung, 2005; Velázquez, 2007; Chapagain and Hoekstra, 2008). Some of the authors promoting this perspective suggest that the notion of virtual water is analogous to the economic concept of comparative advantage, which is a core principle of international trade theory (Allan 2003; Lant, 2003). However, that perspective is not fully accurate.

Comparative advantage requires consideration of the opportunity costs of production for each trading partner. The opportunity costs will depend on resource endowments and the technology of production in each setting. The virtual water perspective considers only a country's water endowment. There is no consideration of technology and no comparison of the opportunity costs of production within or across trading partners. At its best, virtual water might be described as an application of absolute advantage, which is not a sufficient criterion for determining optimal trading strategies (Wichelns, 2004; Wichelns 2011). Absolute advantage neglects consideration of opportunity costs, which must be considered to identify the strategy that maximizes the sum of net benefits from international trade.

The suggestion that water-short countries should not produce and export water-intensive crops could encourage policy makers to promote production and trade strategies that reduce social net benefits. Several water-short countries, such as Israel, Jordan, and Australia, produce and trade water-intensive products. Those activities generate substantial revenues for the producers, while enhancing the portfolio of goods and services available in both the exporting and importing countries.

The conceptual inadequacy of the virtual water perspective is reinforced by empirical analysis of international trade. Kumar and Singh (2005) analyze data describing water availability and international trade for 146 countries. They find that observed trading patterns are not consistent with those predicted by the virtual water perspective. Some water-abundant countries import food, while some water-scarce countries export food. The authors conclude that relative land endowments, access to arable land, and a water storage in the soil profile would be more helpful than water endowments in explaining the observed variation in international trade patterns.

De Fraiture et al. (2004) also find limited empirical support for the virtual water perspective. They caution against inferring that international trade will be helpful in mitigating global water scarcity, in part, because political and economic considerations can have greater influence than water scarcity in determining national trading strategies. Lopez-Gunn and Llamas (2008) also observe that international trade in food is driven largely by factors other than water.

Wichelns (2010a, 2010b) examines the estimates of virtual water imports and exports prepared for 77 countries by Chapagain and Hoekstra (2004). He concludes that the amount of arable land per person in a country is a better descriptor of international trade patterns than is the amount of renewable water resources available, per person or per hectare. A country's arable land endowment is not a sufficient predictor of trade patterns, but it is a better descriptor of trade in crop and livestock products than a country's water endowment.

Ramirez-Vallejo and Rogers (2004) examine empirical information in the context of the Heckscher-Ohlin model of international trade, which suggests that countries trade on the basis of the relative abundance of factors of production. This is a somewhat less restrictive criterion than the theory of comparative advantage. Yet the authors find little empirical support for predicting trade patterns on the basis of national water endowments. Rather, they find that variables such as average income, population, irrigated area, and the amount of value added in agriculture are helpful in explaining the observed variation in the trading of agricultural commodities.

Guan and Hubacek (2007) examine the current movement of agricultural products between northern and southern China, with the goal of determining whether or not the data reflect implementation of a virtual water trading strategy. Their null hypothesis is that water-scarce northern China will import water-intensive goods and export goods requiring less water in production, while water-abundant southern China will operate in reverse. The data do not support the "virtual water hypothesis." Water-scarce northern China exports many water-intensive goods and services, while water-abundant southern China imports water-intensive goods. The authors suggest that several factors influencing agricultural input use and productivity—water price, labor availability, and soil and land quality— might be responsible for the results they have observed. These factors are among those that help determine opportunity costs and comparative advantages.

In several of the examples cited here, the authors suggest that arable land, irrigated area, or water stored in the soil profile is a more helpful indicator of international trading strategies than is the estimate of a country's water endowment. Water resources are certainly considered when preparing estimates of arable land, irrigated area, and soil moisture. Yet by including other factors, such as land and irrigation investments, the alternative indicators have greater predictive usefulness than estimates of water endowments.

### Regarding Water Savings and Losses

Several authors have suggested that countries save or lose water when they engage in international trade. The authors obtain their estimates of water savings and losses by examining water requirements in the importing and exporting countries (Yang et al., 2006; Hoekstra and Chapagain, 2007b; Yang and Zehnder, 2007). For example, a country importing crops that would have required 12 million ML of water to produce, domestically, is considered to have saved those 12 million ML. If the exporting country used only 10 million ML to produce the crops, then the global water savings attributed to the transaction is considered to be 2 million ML. Some authors sum their estimates of such water savings to determine the volume of water saved globally through the merits of "virtual water trade" (Oki and Kanae, 2004).

Several examples illustrate the estimation of water savings and losses by authors using the notion of virtual water as an analytical framework. Mekonnen and Hoekstra (2010) suggest that by importing 960,000 tons of wheat each year from France, Morocco saves the 3.77 million ML of water that would be required to produce the wheat domestically. As the wheat is produced in France using only 0.6 million ML per year, the annual wheat trade between Morocco and France contributes 3.17 million ML to the authors' estimate of annual global water savings due to international trade in wheat products, which is 65 million ML.

Chapagain et al. (2006a) provide similar analysis of virtual water volumes moving between countries as they engage in the trade of agricultural products. They suggest, for example, that Mexico saves 1.06 million ML of water annually by importing 488,000 tons of husked rice from the United States. The estimated global water savings due to the trade is 0.44 million ML, as the amount of water required to produce the rice in the United States is 0.62 million ML.

The authors consider also that countries lose water when they export crops. For example, Thailand is said to lose an estimated 2.27 million ML of water each year through its export of 416,000 tons of broken rice to Indonesia (Chapagain et al., 2006a). There is an estimated global water loss of 0.98 million ML, as well, given that the rice could have been produced in Indonesia using only 1.29 million ML. The estimates of virtual water content underlying this result are 5.455 ML per ton of rice in Thailand and 3.103 ML per ton in Indonesia.

Some of the estimated national savings and losses of water are quite large. Japan is said to save an estimated 94 million ML per year through its imports of crop and livestock products, while Mexico saves an estimated 65 million ML per year, and Algeria saves an estimated 45 million ML per year. A careful review of resource endowments and production opportunities in selected countries reveals that such estimates of water savings and losses are not appropriate.

In many countries, the amount of arable land is limited, while water is relatively abundant. It seems reasonable to ask if such countries actually save water when they import agricultural products. The industrialized coun-tries of Japan, Germany, and the Republic of Korea are among those with the largest estimated annual net water savings made possible through international trade in agricultural products (Chapagain et al., 2006a). In each of those countries, there are fewer than 0.15 ha of arable land per person, and less than 5 percent of the population is involved in agriculture (Table 6.1). Each country has a viable agricultural sector, but the sector accounts for no more than 3 percent of gross domestic product. None of the countries has been self-sufficient in food production for many years. Each of the countries has comparative advantages in other productive activities.

Similar observations can be made regarding other countries for which the estimated net water savings due to international trade are quite large. All of the countries in Table 6.1, except Russia, have fewer than 0.30 ha of arable land per person. In six of the countries, the proportion of residents involved in agriculture is smaller than 10 percent. In each of those countries, agriculture accounts for less than 5 percent of gross domestic product. In such settings, it seems misplaced to consider that countries save water through their imports of food and other products.

It is not correct, for example, to suggest that Japan saves 94 million ML of water each year. Japan does not have the option of using that volume of water to expand crop and livestock production. Japan and many other industrialized countries must import crop and livestock products to sustain economic development. With small amounts of arable land per person, and largely urbanized societies, Japan and other countries will not be return-ing to their agrarian past. Hence, it is not helpful to suggest that Japan saves water by engaging in international trade. The numerical estimate of Japan's water savings is not meaningful and it has no policy relevance.

### Considering Impacts on Livelihoods

In addition to lacking consideration of opportunity costs, the virtual water perspective neglects consideration of the impacts of production and trade

*Table 6.1* Estimated annual national net water savings, arable land, and the importance of agriculture in selected countries

| Country | Net national water savings (bcm/year) | Arable land per person (ha per person) | Agriculture as proportion of gross domestic product (%) | Proportion of population involved in agriculture (%) |
|---|---|---|---|---|
| Japan | 94 | 0.03 | 1.6 | 2.1 |
| Mexico | 65 | 0.22 | 4.3 | 17.9 |
| Italy | 59 | 0.12 | 1.8 | 3.3 |
| China | 56 | 0.10 | 10.6 | 60.8 |
| Algeria | 45 | 0.21 | 8.4 | 20.9 |
| Russia | 41 | 0.87 | 4.7 | 8.0 |
| Iran | 37 | 0.22 | 10.9 | 21.5 |
| Germany | 34 | 0.14 | 0.9 | 1.6 |
| Republic of Korea | 34 | 0.03 | 3.0 | 4.6 |
| United Kingdom | 33 | 0.10 | 1.2 | 1.5 |
| Morocco | 27 | 0.25 | 19.2 | 25.8 |

Notes: bcm, billions of cubic meters; ha, hectares.

Sources: The list of countries and the estimates of net national water savings resulting from international trade in agricultural products during 1997 to 2001 are from Chapagain et al. (2006a). Estimates of arable land per person and the proportion of the population involved in agriculture are from the Food and Agriculture Organization of the United Nations (FAO) (http://faostat.fao.org). Estimates of agriculture as a proportion of gross domestic product are from the Central Intelligence Agency's (CIA) World Factbook (www.cia.gov).

on the livelihoods of individuals and the vibrancy of communities engaged in agriculture. Proposals to rearrange international trading patterns based only on consideration of water endowments could impose substantial harm on individuals who earn their living in agriculture, particularly in poor countries. For example, suppose an international trade authority proposed that Vietnam reduce or eliminate rice exports, because rice production has a large water footprint. Such a program could reduce livelihood opportunities substantially for thousands of farm families, even though Vietnam might have a comparative advantage in rice production. In addition, such a proposal would not account for the appropriateness of producing rice in monsoonal climates or the very long history and cultural attachment to rice production in Vietnam.

## 6.3   Inadequacies of Water Footprints

This brief review of literature suggests that the notions of virtual water and water footprints might lack the information and insight required to determine optimal strategies regarding water resources. The notions might be helpful in raising awareness, but they might not be policy relevant.

## Too Little Information

Estimates of water footprints contain too little information to enhance understanding of water resource issues and inform policy makers of the best ways to improve the social net benefits of water allocation and use. Water footprints consider only the volume of water used in production, without considering other inputs or the opportunity costs of any inputs. Comparing two water footprints, across locations or over time, is not a helpful exercise if one does not have information regarding water scarcity conditions, the opportunity costs of water, and water's role in supporting livelihoods in the settings in which the water footprints are calculated.

Water volumes, alone, are not sufficient indicators of the benefits or costs of water use in any setting. The benefits and costs are functions of complex interactions involving physical, economic, and social dimensions that are not contained or reflected in estimates of water footprints. The water footprint of coffee might be 140ml per cup, but that estimate provides no insight regarding the opportunity cost of water in the region where the coffee is produced, or the beneficial livelihood impacts for those people engaged in coffee production. Coffee produced in a country with abundant water might place no pressure on water supplies. Yet the activity might provide livelihoods to many residents with few alternative sources of employment. Such aspects of water allocation decisions are not reflected in estimates of water footprints.

It is critical to consider both the opportunity cost of water (its scarcity value) and the opportunity cost of labor (alternative employment options) when evaluating policies that impact the allocation and use of water and other productive inputs. The water footprint of a coconut might be 2,500 liters per kg, but most coconuts are produced in humid regions with abundant water supplies. In such settings, the opportunity cost of much of the water used in coconut production is not substantial, and local residents might have few alternatives to earning their livelihoods in the production and processing of coconuts.

Water footprints lack also any information regarding the incremental benefits made possible by allocating and using water in production and consumption activities. Water is an essential ingredient of both the supply and demand components of local and international food markets. It is not possible to determine optimal water allocations or the best possible uses of water in society by considering only the volumes of water consumed in selected activities.

## Inappropriate International Linkages

Several authors have calculated the internal and external water footprints of selected countries (Hoekstra and Chapagain, 2007b; Fader et al., 2011). Some of those authors suggest that consumers of one country are partly responsible for water resource issues in another country, as their external water footprints place demand pressures on distant water resources. Van Oel et al. (2009) report that 89 percent of the estimated per capita water footprint for the Netherlands is external. Furthermore, they suggest that the external water footprint imposes negative impacts on the countries from which the Dutch import goods and services. Chapagain et al. (2006b) suggest that Japan's demand for cotton exerts pressure on water resources in Pakistan, China, and India. The authors suggest also that European consumers, through their consumption of cotton products, contribute indirectly to about 20 percent of the desiccation of the Aral Sea. In aggregate, about "half of the water problems in the world related to cotton growth and processing can be attributed to foreign demand for cotton products" (p. 201).

The suggestion that producers and consumers in one country are responsible for resource issues in another is compelling in many cases, but is not always helpful. Consumers in faraway lands certainly can influence production and marketing decisions through their collective action as purchasers of imported goods and services. Successful examples of consumer campaigns include efforts to boycott products assembled in factories using child labor, and efforts to end the harvest of tuna using methods that are harmful to dolphins (Teisl et al., 2002; Brown, 2005; Basu et al., 2006; Baird and Quastel, 2011). In each of those cases, consumers have determined appropriately that any amount of child labor is undesirable, and that the loss of any dolphin is unacceptable. Such clear-cut criteria do not apply in the case of water use.

The water footprint of coconut production in Sri Lanka or Malaysia should not cause concern among consumers of canned coconut meat in New York. There is sufficient water in Sri Lanka and Malaysia in most seasons to support coconut production, without harming the environment or restricting the supply of water available to other users. The opportunity cost of water in many areas of very humid countries is considerably small in most seasons. A global campaign to reduce water footprints would not be sensible, given that water use in many settings is not causing environmental harm or limiting economic development.

### Incorrect Analogy to Carbon and Ecological Footprints

Many readers of the literature might develop the impression that water footprints are analogous to ecological and carbon footprints (Hoekstra, 2009; Ercin and Hoekstra, 2012). Yet the technical characteristics of each of the footprints are quite different. Ecological and carbon footprints attempt to describe the global implications of human activities in any local or regional setting (Moran et al. 2008; Kitzes and Wackernagel 2009; Peters 2010). Economic activities place demands on the planet's productive and assimilative capacity (ecological footprints), while releasing waste materials that modify and degrade the atmosphere (carbon footprints). In each case, the activities generate impacts that can be summed and compared in a meaningful fashion across locations. For example, the carbon emitted by automobiles in New York has the same impact on the atmosphere as the carbon emitted by cars in Beijing. Hence, there is some logic in comparing and summing the carbon footprints of residents in those cities.

In concept, a reduction in carbon footprints in any country will reduce global pressure on the atmosphere. The same is not true of water footprints. Residents of humid countries generally will have larger water footprints than residents of arid countries. Yet those large water footprints might be sustainable and they might impose no harm on the environment. There might be no social gain from efforts to reduce water footprints in such settings. Rather, there might be social losses if such efforts reduce the demand for goods and services in industries that provide gainful employment, enhance livelihoods, or contribute to international trade.

Ecological footprints are intended to represent the demands that production and consumption activities place on the earth's potential to provide productive resources and assimilate waste materials. Such footprints can be calculated for individuals, communities, and countries. The notion motivating ecological footprint analysis is that the earth's productive potential is limited. If the demands we place on the earth through our production and consumption activities exceed the planet's productive and assimilative capacity, sustainability will not be achieved.

Those who estimate ecological footprints express the demands and supply of productive and assimilative capacity in terms of a common metric, to enable summation and comparison of footprints across regions and countries (Wackernagel, 2009). In particular, the authors express demand and supply in terms of "global hectares," which are intended to reflect the areas of land and sea required to support production and consumption activities, and assimilate waste materials. The estimates of global hectares are used for two purposes: 1) to compare the demand pressures imposed by individuals and communities, and 2) to determine whether the sum of demand pressures is greater than the available supply of the earth's productive and assimilative capacity.

Some authors (Moran et al., 2008; Kestemont et al., 2011) have shown that the ecological footprints of residents in wealthy countries exceed those of residents in poor countries. Others (Ewing et al., 2010) have shown that the sum of ecological footprints has been greater than earth's capacity since the middle 1970s, and the gap between the aggregate ecological footprint and the earth's capacity is increasing. Such analyses, if accurate and appropriate, would provide guidance regarding whether or not the global population is using the earth's resources in a sustainable manner. In addition, the analysis enables one to examine opportunities for reducing global demand pressures, by noting regions in which the ecological footprints are substantially higher than productive capacity. In theory, reducing an ecological footprint in one region will contribute to reducing the sum of ecological footprints across regions or countries.

The comparison and summary properties of ecological footprints are not shared by water footprints for two reasons: 1) water scarcity issues, which involve the imbalance between supply and demand, are regional and local, rather than international, and 2) estimates of water footprints contain no information describing impacts. We should expect the water footprints of consumption and production activities in humid countries to be larger than those in arid countries, all else equal. The water footprint of banana production likely will be greater in Sri Lanka than in Jordan. Yet the larger water footprint might not be causing any harm to local residents or the international community. Reducing the water footprints of residents in Amsterdam will not enhance water availability in Amman. Comparing or summing the water footprints of consumers in both cities will not provide policy relevant information. The notion of a water footprint is too narrow in scope to guide public officials toward appropriate policy decisions.

### Inadequate Consideration of Costs and Benefits

Reducing water footprints is not always a desirable objective. Water footprints consider only water volumes, which are not sufficient indicators of the benefits or costs of water use in any setting. The benefits and costs are functions of complex interactions involving physical, economic, and social dimensions that are not contained or reflected in estimates of water footprints. For example, in many humid areas, where water is not scarce, the costs of reducing water deliveries to agriculture might exceed the benefits. The expenditures on labor, energy, and equipment required to improve irrigation management might exceed the incremental value of reducing irrigation diversions, particularly in regions where surface runoff and deep percolation are useful resources.

Public officials must consider an array of questions pertaining to incremental benefits and costs, before reaching decisions regarding water resource allocation and use. For example, they must consider the scarcity costs and environmental implications of non-water inputs in the production of goods and services. Examples include land, labor, energy, fertilizer, pesticides, and machinery. Farm-level decisions regarding water use on farms can influence the amounts of these other inputs used, as well. Efforts to reduce water footprints can result in greater use of electricity or farm machinery, thus increasing any off-farm impacts associated with those inputs.

One must also consider very carefully the inherent water scarcity conditions. As noted above, it might be unwise to reduce water footprints in areas where water is not scarce, particularly if there are notable direct or indirect costs involved in such efforts. Water and other natural resources are critical inputs in household production functions for much of humanity.

Efforts to reduce water footprints regionally or as part of a national strategy, can have severe implications on employment opportunities in agriculture and on household level access to water resources. Public officials must consider the potential impacts of their initiatives regarding water resources on food security and livelihoods, rather than simply attempting to reduce a volumetric measure of water use.

In addition to its role as a critical input in crop production, water is required for many activities at the household level (Smits et al., 2010; van Koppen and Smits, 2010). In many areas of developing countries, individual and household water footprints are too small, rather than too large. Yet the manner in which water footprints generally are presented in the literature seems to suggest that smaller is better, and that consumers and producers everywhere should endeavor to reduce their water footprints. A broader view that embraces the many benefits of water use would be more appropriate and more helpful, particularly when discussing public policies.

## 6.4  Summing Up

The notions of virtual water and water footprints have gained the attention of many water resource scholars and practitioners. Many authors apply these notions as analytical tools to describe international trade or to propose that consumers in one region are responsible for environmental issues in another. Yet there is no conceptual model or an established empirical basis supporting such analysis. Virtual water and water footprints can raise awareness of water issues by describing the amounts of water required to produce goods and services, but the notions do not contain sufficient information to determine smart public policies or to guide discussions regarding international trade.

The notion of virtual water has been taken far beyond the intent for which it was first introduced—that of describing the amount of water required to produce the food imported by countries with limited arable land. Indeed, many arid countries must import large amounts of grain and other commodities to ensure food security. Yet the importing countries have no reason to inquire about the amount of water used to produce the grain they purchase. They have reason to care very much about price, availability, and quality, but little reason to be concerned with the amount of water used in production. Thus, the notion of virtual water is not helpful in describing how international trade actually takes place. Nor should the notion be used to design alternative trading patterns. There are many additional issues to consider when crafting policy measures.

Water footprints also fall short as an analytical construct. They lack sufficient information to support policy analysis or to determine optimal decisions by consumers and firms. Just as virtual water is silent on the issue of opportunity costs, water footprints also neglect information describing water scarcity conditions, implications for livelihoods, and the beneficial aspects of water use in any setting. Even in water scarce areas, residents, firms, and public officials will be concerned about energy resources, land use, other inputs, and the implications of water allocation and use on livelihoods.

Firms and consumers certainly can gain information and increase their awareness by calculating water footprints. There is no inherent harm in such calculations, although firms and consumers might wish to take a broader perspective regarding their activities by considering also other inputs and livelihood impacts. Many firms likely already consider the amount of water they use in production, particularly if they operate in regions where water is scarce or priced by volume. The firms might not use the language of methodology of water footprints, but understanding input use is certainly in their interest. Thus, one might conclude that water footprints are not harmful when calculated and applied in private settings. The potential harm arises when water footprints are considered in a policy context.

Designing policies based on water footprints might lead to decisions that move society further away from desirable outcomes. An estimated water footprint might describe the volume of water used in production, but it does not describe the amounts of other inputs, or the opportunity cost of any input used to produce goods and services. Producers required to reduce their water footprints might increase their use of energy or machinery, even if the opportunity costs of those resources are higher than the opportunity cost of water in a given location. Consumers encouraged to reduce their water footprints might choose products with smaller footprints depicted on product labels. Yet their choices might displace low-wage laborers in a region with few alternative employment opportunities.

In sum, virtual water and water footprints are compelling notions, but they have limited relevance in discussions of important policy questions. International trade is complex, and involves many issues that are not captured in calculations or depictions of virtual water. Similarly, water use and allocation involve many more issues and implications than are reflected in estimates of water footprints.

# References

Aldaya, M. M., Martínez-Santos, P., and Llamas, M. R. 2010. Incorporating the water footprint and virtual water into policy: Reflections from the Mancha Occidental Region, Spain. *Water Resources Management* 24(5): 941–958.

Allan, J. A. 1996. Water use and development in arid regions: Environment, economic development and water resource politics and policy. *Review of European Community International Environmental Law* 5(2): 107–115.

Allan, J. A. 2002. Hydro-peace in the Middle East: Why no water wars? A case study of the Jordan River Basin. *SAIS Review* 22(2): 255–272.

Allan, J. A. 2003. Virtual water—the water, food, and trade nexus: Useful concept or misleading metaphor? *Water International* 28(1): 4–11.

Baird, I. G., and Quastel, N., 2011. Dolphin-safe tuna from California to Thailand: Localisms in environmental certification of global commodity networks. *Annals of the Association of American Geographers* 101(2): 337–355.

Basu, A.K., Chau, N. H., Grote, U., 2006. Guaranteed manufactured without child labor: The economics of consumer boycotts, social labeling and trade sanctions. *Review of Development Economics* 10(3): 466–491.

Brown, J. 2005. An account of the dolphin-safe tuna issue in the UK. *Marine Policy* 29(1): 39–46.

Chapagain, A. K., and Hoekstra, A. Y. 2004. *Water Footprints of Nations*, Volume 2: Appendix. Value of Water Research Report Series number 16. Delft, Netherlands: UNESCO-IHE Institute for Water Education.

Chapagain, A. K. and Hoekstra, A. Y. 2008. The global component of freshwater demand and supply: An assessment of virtual water flows between nations as a result of trade in agricultural and industrial products. *Water International* 33(1): 19–32.

Chapagain, A. K., Hoekstra, A. Y., and Savenije, H. H. G. 2006a. Water saving through international trade of agricultural products. *Hydrology and Earth System Science* 10(3): 455–468.

Chapagain, A. K., Hoekstra, A. Y., Savenije, H. H. G., and Gautam, R. 2006b. The water footprint of cotton consumption: An assessment of the impact of worldwide consumption of cotton products on the water resources in the cotton producing countries. *Ecological Economics* 60(1): 186–203.

de Fraiture, C., Cai, X., Amarasinghe, U., Rosegrant, M., and Molden, D. 2004. *Does International Cereal Trade Save Water? The impact of virtual water trade on global water use.* Comprehensive Assessment Research Report 4. Colombo, Sri Lanka: International Water Management Institute.

Ercin, A. E., and Hoekstra, A. Y. 2012. Carbon and water footprints: Concepts, methodologies and policy responses. UNESCO, Paris: Side Publications Series: 04.

Ewing, B., Reed, A., Galli, A., Kitzes, J., and Wackernagel, M. 2010. *Calculation Methodology for the National Footprint Accounts*, 2010 edition. Oakland, CA: Global Footprint Network.

Fader, M., Gerten, D., Thammer, M., Heinke, J., Lotze-Campen, H., Lucht, W., and Cramer, W. 2011. Internal and external green–blue and related water and land savings through trade. *Hydrology and Earth System Sciences* 15(5): 1641–1660.

Government of India, 2012. *National Water Policy (2012).* Available at: http://mowr.gov.in/index1.asp?linkid=201&langid=1 (accessed August 5, 2013).

Guan, D., and Hubacek, K. 2007. Assessment of regional trade and virtual water flows in China. *Ecological Economics* 61(1): 159–170.

Hoekstra, A. Y. 2009. Human appropriation of natural capital: A comparison of ecological footprint and water footprint analysis. *Ecological Economics* 68(7): 1963–1974.

Hoekstra, A. Y. 2013. *The Water Footprint of Modern Consumer Society.* New York: Routledge.

Hoekstra, A. Y., and Chapagain, A. K. 2007a. Water footprints of nations: Water use by people as a function of their consumption pattern. *Water Resources Management* 21(1): 35–48.

Hoekstra, A. Y., and Chapagain, A. K. 2007b. The water footprints of Morocco and the Netherlands: Global water use as a result of domestic consumption of agricultural commodities. *Ecological Economics* 64(1): 143–151.

Hoekstra, A. Y., and Hung, P. Q. 2002. *Virtual Water Trade: A quantification of virtual water flows between nations in relation to international crop trade.* Value of Water Research Report Series No. 11. Delft, Netherlands: IHE.

Hoekstra, A. Y., and Hung, P. Q. 2005. Globalization of water resources: International virtual water flows in relation to crop trade. *Global Environmental Change* 15(1): 45–46.

Kestemont, B., Frendo, L., and Zaccai, E. 2011. Indicators of the impacts of development on environment: A comparison of Africa and Europe. *Ecological Indicators* 11(3): 848–856.

Kitzes, J., and Wackernagel, M. 2009. Answers to common questions in ecological footprint accounting. *Ecological Indicators* 9(4): 812–817.

Kumar, M.D., and Singh, O.P. 2005. Virtual water in global food and water policy making: Is there a need for rethinking. *Water Resources Management* 19(6): 759–789.

Lant, C. 2003. Commentary. *Water International* 28(1): 113–115.

Lopez-Gunn, E., and Llamas, M. R. 2008. Re-thinking water scarcity: Can science and technology solve the global water crisis? *Natural Resources Forum* 32(3): 228–238.

Ma, J., Hoekstra, A. Y., Wang, H., Chapagain, A. K., Wang, D., 2006. Virtual versus real water transfers within China. *Philosophical Transactions of the Royal Society B* 361: 835–842.

Mekonnen, M. M., and Hoekstra, A. Y. 2010. A global and high-resolution assessment of the green, blue and grey water footprint of wheat. *Hydrology and Earth System Science* 14(7): 1259–1276.

Moran, D. D., Wackernagel, M., Kitzes, J. A., Goldfinger, S. H., Boutaud, A. 2008. Measuring sustainable development—nation by nation. *Ecological Economics* 64(3): 470–474.

Oki, T., and Kanae, S. 2004. Virtual water trade and world water resources. *Water Science and Technology* 49(7): 203–209.

Peters, G. P. 2010. Carbon footprints and embodied carbon at multiple scales. *Current Opinion in Environmental Sustainability* 2(4): 245–250.

Ramirez-Vallejo, J., and Rogers, P. 2004. Virtual water flows and trade liberalization. *Water Science and Technology* 49(7): 25–32.

SABMiller and WWF-UK, 2009. *Water Footprinting: Identifying and addressing water risks in the value chain.* Woking: SABMiller, and Godalming, WWF-UK.

Smits, S., van Koppen, B., Moriarty, P., and Butterworth, J. 2010. Multiple-use services as an alternative to rural water supply services: A characterization of the approach. *Water Alternatives* 3(1): 102–121.

Teisl, M. F., Roe, B., and Hicks, R. L. 2002. Can eco-labels tune a market? Evidence from dolphin-safe labeling. *Journal of Environmental Economics and Management* 43(3): 339–359.

Van Koppen, B., and Smits, S. 2010. Multiple-use water services: Climbing the water ladder. *Waterlines* 29(1): 5–20.

Van Oel, P. R., Mekonnen, M. M., Hoekstra, A. Y. 2009. The external water footprint of the Netherlands: Geographically-explicit quantification and impact assessment. *Ecological Economics* 69(1): 82–92.

Velázquez, E. 2007. Water trade in Andalusia. Virtual water: An alternative way to manage water use. *Ecological Economics* 63(1): 201–208.

Wackernagel, M. 2009. Introduction: Methodological advancements in footprint analysis. *Ecological Economics* 68(7): 1925–1927.

Wichelns, D. 2004. The policy relevance of virtual water can be enhanced by considering comparative advantages. *Agricultural Water Management* 66(1): 49–63.

Wichelns, D. 2010a. Virtual water: A helpful perspective, but not a sufficient policy criterion. *Water Resources Management* 24(10): 2203–2219.

Wichelns, D. 2010b. Virtual water and water footprints offer limited insight regarding important policy questions. *International Journal of Water Resources Development* 26(4): 639–651.

Wichelns, D. 2011. Virtual water and water footprints: Compelling notions, but notably flawed. *GAIA Journal* 20(3), 171–178.

Witmer, M. C. H., and Cleij, P. 2012. *Water Footprint: Useful for sustainability policies?* PBL Netherlands Environmental Agency, PBL Publication number: 500007001.

Yang, H., Wang, L., Abbaspour, K. C., and Zehnder, A. J. B. 2006. Virtual water trade: An assessment of water use efficiency in the international food trade. *Hydrology and Earth System Science* 10(3): 443–454.

Yang, H., and Zehnder, A. J. B. 2002. Water scarcity and food import: A case study for southern Mediterranean countries. *World Development* 30(8): 1413–1430.

Yang, H., and Zehnder, A. J. B. 2007. "Virtual water": An unfolding concept in integrated water resources management. *Water Resources Research* 43(12): 1–10.

# 7 Green and Blue Water

*Aditya Sood, Sanmugam A. Prathapar,*
*and Vladimir Smakhtin*

## 7.1 Introduction

Researchers are always looking for new approaches to tackle obstinate problems. While in some cases these approaches represent fundamentally new methods of addressing an issue, in other cases they simply constitute a changed perception for the same one. If the change in perception leads to a complete change in societal view of an issue, it is a paradigm shift (Kuhn, 1962). One such shift relates to water management for agricultural production, which is critical to development and will remain so as human population increases and diets shift toward more water demanding food. A sustained effort in improving water and crop productivity in recent decades—exemplified by the Green Revolution of the 1960s and 70s—has allowed food production to outpace human demand (FAO, 1996). However, the point has been reached when low hanging fruits, associated with raising productivity in irrigated areas, have largely been exhausted. As a result, recent discourse is moving toward improving productivity in rainfed agriculture, especially in the African continent where agriculture is predominantly rainfed and productivity is low. To aid in this effort, colors have been utilized to highlight the difference between water management for rainfed versus irrigated agriculture.

Traditionally, in water resource management, colors have been used to define water streams in the realm of domestic wastewater management (Otterpohl et al., 1999; Otterpohl, 2002; Gaulke, 2006; Bester et al., 2008; Daigger, 2009). Four colors are predominantly used to define domestic wastewater. Yellow water is the urine component and the brown water is the water mixed with human feces (Linder, 2007), i.e., toilet wastewater minus yellow water. Black water is a technical term used to define toilet wastewater, i.e., combination of yellow and brown water. Finally, grey water is all of the remaining domestic wastewater, i.e., domestic wastewater without black water.

More recently, the two colors increasingly used in discourse in agricultural water management are green and blue. The role of green water in agriculture has received growing recognition for its ability to foster food security and development outcomes (e.g., Falkenmark, 1995; Falkenmark and Rockström, 2006; Rost et al., 2008; Aruna, 2009; Rockström et al., 2009). Falkenmark and Rockström (2006, p. 129) note that "conventional water-resource perceptions are incomplete" and require widening to include green water. Aruna (2009) notes that consideration of green water in water management is essential for alleviating future food crises. Rockström et al. (2009) highlight how consideration of green water is critical in coping with climate change and show that the countries that might be water scarce in terms of blue water could still be food secure if green water is managed well. A common denominator in most of these studies is advocacy for a greater role for rainfed agriculture.

But does use of green water constitute a paradigm shift or simply reflect application of a new name to an old concept? And has the addition of these two water colors improved understanding of water issues in agriculture or created more confusion? While Prathapar (2012) and Batchelor (2013) posed questions of whether "coloring" water has improved or dumbed-down the water management sector, the aggregate level of scrutiny applied to these new entrants to water management dictionary has not been voluminous. To clarify the benefits and additional insights achieved through use of the green–blue water distinction, this chapter reviews these two "water colors" in relation to pre-existing concepts, as a means to identifying the value added by these terms. The chapter first looks at the definitions of green and blue water as mentioned in scientific literature (section 7.2) and compares green and blue water concepts with traditional hydrology and water resources management terms and notions, especially in agriculture (sections 7.3 and 7.4 respectively). A discussion and conclusion section then focuses on possible softer benefits of water colors, and factors that drove increasing use of green and blue water (section 7.5).

## 7.2 Green and Blue Water: Background and Definitions

The green water concept was first introduced by Falkenmark (1995) in the context of agricultural productivity in sub-humid and semi-arid regions. Falkenmark describes green water as the rainwater that infiltrates into the root zone of plants, used for biomass production (Falkenmark, 1995), and "invisible water in the soil," which is "often forgotten" in water resource management (Falkenmark, 2008). The water that either runs off from the soil surface or percolates beyond the root zone to form groundwater, by contrast, is called the "blue water." A debate on the precise demarcation between green and blue water has been going on since the terms were introduced (Ringersma et al., 2003). This has spurred attempts by numerous experts to provide definitions of the two concepts (Table 7.1).

Blue water definitions are similar and in essence considered any water that can be withdrawn from a water storage infrastructure. The storage infrastructure can be either surface storage such as lakes, reservoirs, rivers,

Table 7.1 Definitions of green and blue water as used in selected publications

| Study | Green water | Blue water | Traditional terminology for green water | Traditional terminology for blue water |
|---|---|---|---|---|
| Falkenmark, 1995 | The rainwater that infiltrates into the root zone and is used for biomass production | The water that either runs off from the soil surface or percolates beyond the root zone to form groundwater | Infiltration and soil moisture | Flow and recharge |
| Zaag et al., 2002 | The part of rainfall that infiltrates into the root zone and is directly used by plants for biomass production through transpiration | Renewable water that occurs in rivers and aquifers | Infiltration and transpiration | River flow and groundwater recharge |
| Ringersma et al., 2003 | Water resource held in the soil that is available to plants | Measurement of water flow (that is abstracted from rivers, surface storage, or groundwater) | Soil moisture | Water withdrawal |
| Falkenmark (2008) | Invisible water in the soil | Visible water contained in rivers and aquifers | Soil moisture | Rivers and groundwater recharge |
| Karlberg et al., 2009 | Resource: The soil moisture generated from infiltrated rainfall that is available for root water uptake by plants | Resource: The stored runoff in dams, lakes and aquifers. Rivers are blue water flows | Soil moisture (field capacity—wilting point) | Reservoirs, lakes, and groundwater recharge |
| Hoff et al., 2010 | Soil water held in the unsaturated zone, formed by precipitation, and available to plants | Liquid water in rivers, lakes, wetlands, and aquifers, which can be withdrawn | Soil moisture (field capacity—wilting point) | Reservoirs, lakes, and groundwater recharge |
| Fader et al., 2011 | Precipitation stored in the soil and evapotranspired on cropland | Water taken from rivers, reservoirs, lakes, and aquifers and used for irrigation | ET | Water withdrawal |
| Mekonnen and Hoekstra, 2011 | Green water footprint: The rainwater consumed | Blue water footprint: Surface and groundwater consumed | ET | Surface water consumption |
| Xu, 2013 | Water that comes from precipitation and is stored in soil and then consumed by vegetation | Resource in aquifers, lakes, wetlands, and impoundments | Transpiration | Reservoirs, lakes, and groundwater recharge |

wetlands etc. or subsurface storages such as aquifers. There are three pathways in which blue water is said to be "generated": i) runoff from the soil surface that reaches the surface storage structures, ii) water flowing laterally within the soil layers and reaching the surface water storage structures, or iii) water that moves vertically through the soil layers that recharges aquifers and can be withdrawn later.

While definitions of green water also look similar upon casual reading, discrepancies emerge on closer inspection (Table 7.1). One group of experts appears to consider green water to include only transpired water. Falkenmark (1995) and Xu (2013), for example, consider transpiration (i.e. infiltrated water used for biomass production) as green water. Similarly, Savenije (1998) only considers water that is transpired through plants as green water, and Zaag et al. (2002) consider infiltration along with transpiration by plants as green water. To differentiate between evaporation and transpiration, Savenije (1998) in fact introduced a new color—"white water"—to define the rainwater that is directly evaporated to the atmosphere and does not participate in the photosynthesis of the plants. It could be either from the bare soil or from land cover.

Another group of authors have a broader interpretation of green water (Table 7.1). Rockström (1999) suggests including evaporation from open water and plant interception as part of green water. Ringersma et al. (2003) consider green water as the water in the soil that is available for plants, but lump evaporation and transpiration together. Similarly, Fader et al. (2011) consider actual ET as green water and do not differentiate between evaporation and transpiration. Finally, Mekonnen and Hoekstra (2011) describe green water footprint simply as rainwater consumed. If the water consumed by a crop (or a derived crop product) comes directly from rainwater, it forms part of green water footprint; if it comes from surface and ground water, it is considered as part of blue water footprint.

Yet another group of authors look at green water in terms of its potential (Table 7.1). Karlberg et al. (2009) talk about the potential of green water, i.e., soil moisture available but not necessarily consumed by plants. Karlberg et al. (2009) also explain that irrigation when applied to fields helps increase green water in the soil and hence leads to "blue to green water redirection." Hoff et al. (2010) consider green water as the precipitation held in an unsaturated zone of soil and available for plants—which implies that they too refer to the potential and not actual ET. They suggest rainwater harvesting[1] as being at the "interface" of blue and green water. This raises issues about interexchange between green and blue water resources. Rainwater stored at on-site small ponds is considered as blue water, for example, until it infiltrates the soil to augment soil moisture—thereby converting to green water.

At least two points emerge from the above review of definitions. A first point is that blue water definitions are fairly consistent. A second point is that variation in interpretation is evidenced across definitions of green water.

To some, green water is the soil moisture that is transpired; to others, it is soil moisture that is evapotranspired; and yet to others, it is any moisture in the soil that can be eventually used by the plants.

## 7.3 Hydrology and Water Colors

To further clarify the complexities involved in defining green and blue water, it is important to clearly situate these concepts within the hydrological cycle at a basin scale (Figure 7.1a). There are two natural ways in which water enters a river basin. It can either enter as precipitation (in the form of snow or rainfall) or through groundwater flow if the groundwater boundaries do not match the watershed boundaries. Precipitation either falls directly on the bare soil or on vegetation. Some of the precipitation that falls on the vegetation is retained there and later evaporates back to the atmosphere— this is interception. Some scientists consider the interception as part of the green water, while others do not.

Rain that falls on vegetation and eventually reaches the soil surface is known as "throughfall" (Figure 7.1a). Some of the rainwater that reaches the surface of the soil infiltrates into the soil. The rest of the rainwater flows over the land and either collects in surface depressions or flows into lakes and rivers. This process is influenced by many factors such as soil properties (compaction, hydraulic conductivity, texture, existing soil moisture etc.), rainfall (intensity, duration, and amount), topography (slope, exposure), and vegetation.

Rainwater that does not infiltrate is considered blue water. The rainwater that infiltrates in the soil is either retained as soil moisture (in unsaturated zone) or percolates further down and becomes part of the groundwater (saturated zone). The rainwater that is retained in the unsaturated zone of the soil is called green water. And the rainwater that reaches the saturation zone (or groundwater) is called blue water. The blue water that gets stored in shallow depressions or potholes either evaporates back to the atmosphere or slowly seeps into the soil thus becoming green water. Rainwater harvesting techniques that rely on storing rainwater at farm level are analogues to these shallow depressions, and are a good example of "blue to green water redirection."

Figure 7.1b shows hydrological processes in the vadose zone of soil. The vadose zone is the zone between the land surface and the groundwater (where the water is at atmospheric pressure). Water in the vadose zone is held by either adhesion or capillary action and is termed as soil moisture. In terms of water color, it will be considered as green water. Both at the soil surface and interface between vadose zone and groundwater, continuous interactions take place. At the soil surface, soil moisture evaporates due to transfer of energy. This would fall either under green water or white water, depending upon which definition of green water is considered. During rains, the soil moisture is replenished through the soil surface (infiltration).

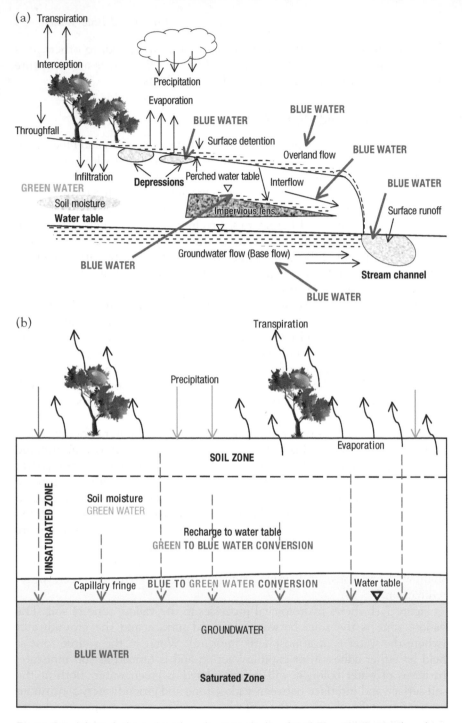

*Figure 7.1* (a) Hydrological cycle at basin scale (modified from FAO, 2013) and (b) Hydrological processes within a vadose zone (modified from United States Geological Survey (USGS), 2013).

At the phreatic surface, where vadose zone and groundwater meet, water is added to the groundwater due to gravitational forces converting green water to blue water. In almost all cases, at the phreatic surface, due to capillary action, a saturated fringe is formed converting blue water to green water. Thus soil may be fully saturated with water well above the groundwater level. In the vadose zone, soil moisture is also lost to the atmosphere through the plants in the form of transpiration. This happens in the root zone of the soil, i.e., the layer up to which the roots of the plant extend. Soil water held within an evaporating plane (the depth of this plane will depend on the depth of the root zone, soil texture and structure) will move to the soil surface against gravity, consumed by roots if present, and evaporate or transpire (green water). Water held below this plane will move very slowly toward the aquifer and become groundwater that can be pumped or transported, thus becoming blue water.

In summary, consideration of green and blue water in the context of the hydrological cycle suggests that both could be described using traditional terminology. Green water can be said to include all or part of soil moisture and interception, depending on which definition of green water is applied. Blue water can be said to include water not retained by the plants or soil. Finally, green to blue water conversion and vice versa occurs at the interface of both types of water.

## 7.4   Water Resource Management, Agriculture, and Water Colors

Water resource management traditionally involves managing water storage and water flows. Two new terms related to green and blue water were introduced—green water flows and blue water flows—in this context (Falkenmark and Rockström, 2006). Karlberg et al. (2009, p. 44) define green water flows as "vapor flows that go back to the atmosphere"—this includes transpiration and evaporation from soil and interception. In other words, green water flows is another name for ET and interception, where transpiration is the productive part and the rest unproductive (Falkenmark and Rockström, 2006). Blue water flows is defined as "the liquid flows of water that recharge groundwater and flow in rivers to lakes, wetlands and ultimately the ocean" (ibid., p. 45). As such, blue water flows is the term for the flow of water in surface and aquifer storage systems.

In the past, water managers may have placed more focus on managing surface water (blue water) and blue water flows than soil moisture (green water). Advocates for increased management of green water (e.g., Falkenmark and Rockström, 2006) may therefore have been highlighting opportunities to exploit a relatively underrepresented piece of the water management puzzle. While this is likely a noble objective, it is worth noting that components of the hydrological cycle are ultimately all connected to each other and an impact on one component affects management of other components. Vidal et al. (2009) indeed suggested that agriculture has never been fully rainfed or fully irrigated—it is somewhere in between. Further, farmers generally consider rainfed and irrigated agriculture together, applying water for irrigation to supplement water from rainfall.

Two systems of agriculture are practiced around the world. One is rainfed, i.e., all the crop requirements are met directly from the rainwater falling on the farm. In this system, the crops extract soil moisture that is retained due to rainfall infiltration or soil moisture that is created due to capillary action, especially in the case of high water table. Crop productivity, i.e., crop produced per hectare, is controlled by rainfall, thus making the system highly vulnerable to natural variability.

The other one is irrigated, i.e., where crop water requirements are met through withdrawal, delivery, and application of water. In reality, most irrigated systems are in fact a combination of irrigation and rainfed agriculture since water available to crops from rainfall is typically supplemented by irrigation, i.e., there is usually some existing soil moisture due to rainfall that is enhanced by irrigation. In general, crop productivity is higher in irrigated systems as compared to rainfed systems because water can be provided to crops in adequate amounts reliably throughout the growing season, thus avoiding undue water stress to the crops (Kijne et al., 2003; Geerts and Raes, 2009).

In the literature cited above, water directly consumed from rainfall in rainfed and irrigated agriculture is termed green water and the water supplied by irrigation in an irrigated system is considered as blue water. Confusion comes when these terms are applied to the case of rainwater harvesting. When a farmer creates a small retention pond to hold rainwater, that farmer is in effect collecting blue water. As this water slowly percolates in the soil, however, it converts to green water. The water may nonetheless eventually become blue water again if it seeps through soil to recharge the aquifer. Similarly, water consumed by crops from the capillary fringe is also blue water converted to green. In reality, crops can only consume soil moisture, which means that any water that is applied to soil, first gets converted to soil moisture before being consumed by crops. It is therefore critical to identify the source of soil moisture if one seeks to determine whether a crop has utilized green or blue water.

Some studies have divided water productivity into "green water productivity" and "blue water productivity" (Rockström et al., 2004; Hamdy, 2008). Water productivity is defined as economic or biophysical output for a unit of water applied or depleted (see Chapter 5 for a full review). As already discussed, since water consumed in most irrigated systems comes from an ad hoc combination of rainfall and withdrawal, it is difficult to disaggregate the agricultural production into rain based and irrigation based. The total water applied to any farm is a sum of rainfall and irrigation (in case of purely rainfed, water from irrigation is zero). Some of the water runs off the field and some percolates down to the water table. At the scale of a farm, this water is lost (although it can be reused downstream). The rest is either consumed by plants or evaporated. The goal of agriculture management at a farm scale is to reduce the loss of water to downstream runoff, to groundwater, or by evaporation.

## 7.5 Discussion and Conclusion

While there has been a notable stream of literature that uses the language of green and blue water, it appears questionable whether these terms have provided insights above and beyond those that are or may be achieved through concepts that were already in use.

Clearly, green and blue water can be described using traditional hydrological terminology. Also, rigid categorization of water as either green or blue is likely to cause more confusion since application of green water definitions typically allow for some "redirection" or "conversion" between green and blue water (and vice versa) as seen, for example, in the case of rainwater harvesting. Confusion is compounded by the fact that there is no common definition of green water (Table 7.1). There are three groups of definitions for green water—water from soil moisture that is evapotranspired, water from soil moisture that is only transpired, and any soil moisture.

It appears that a major driver for the use of "green and blue water" terms is to promote greater focus on rainfed agriculture. Insofar as green and blue water—as well as the rest of rainbow water terminology—further this objective, they may have some merit as science communication concepts. Nonetheless, such terms need to be fully consistent with and reflective of the underpinning science itself. Case in point, green water does not directly equate with rainfed agriculture, nor does blue water equate with irrigation.

It is true that the focus of water resource managers and agriculture engineers has traditionally been more oriented toward surface water and irrigation, and less oriented toward management of soil moisture and groundwater. This narrow interpretation of water management has nonetheless been evolving over the years, as discourse has moved to collective management of surface and groundwater (World Bank, 2005) and rainfed agriculture (CAWMA, 2007). It is being realized now that water resource management needs a holistic approach that manages all the components of a hydrological cycle so as to increase water reliability. This being said, there may still be a niche for an advocacy tool that may help offset the greater attention historically given to withdrawal and surface-water storage in water management.

One possible advantage of introducing the terms green and blue water is that they may be more accessible and appealing than traditional water management terms. "Tapping the potential of green water" may indeed constitute a more compelling and novel advocacy slogan than the more routine "improving water management in rainfed agriculture," particularly to non-technical audiences who will not scratch deeply beyond the surface to test the conceptual soundness of such terms. It has in fact been suggested that management is often about finding a working balance between scientists and those with less technical understandings of water. If this is true, the terms of green and blue water could—in principal—hold value as they may stimulate more interest in water management and be perceived to simplify concepts viewed as esoteric.

The experience of green and blue waters' entry into water management discussions engenders some broader questions about the process of discourse through which new terms enter water management. One indeed wonders whether attempts should be first made to define future agricultural water management problems concisely in traditional hydrological terms and using the existing scientific capital in handling these issues. When clear limitations to old approaches are apparent, it is then valid to explore developing new tools to complement the old ones. Nonetheless, in such cases when utility of old concepts is limited and new terms in water management must be added, it would seem necessary to scrutinize these terms with proper scientific rigor to make sure that they are conceptually sound and add value. The experience of green and blue waters entry into water management discussions engenders some broader questions about *the process* through which new terms enter water management discourse.

## Note

1.  Rainwater harvesting is the practice of capturing and accumulating rainfall locally with the intention of using it later for either human consumption or to recharge groundwater.

## References

Aruna 2009. "Green" and "blue" water provides hope for future food crises. Medindia, available online: www.medindia.net/news/Green-And-Blue-Water-Provides-Hope-For-Future-Food-Crises-50977-1.htm#ixzz2DcEZMXNT

Bester, K., Scholes, L., Wahlberg, C., and McArdell, C. S. 2008. Sources and mass flows of xenobiotics in urban water cycles: An overview on current knowledge and data gaps. *Water, Air, & Soil Pollution: Focus* 8(5–6): 407–423.

CAWMA 2007. *Water for Food, Water for Life: A comprehensive assessment of water management in agriculture.* London: Earthscan, and Colombo: International Water Management Institute.

Batchelor, Charles (ICID member). 2013. Does use of the blue–green–grey water concept improve or dumb-down water management? UEA Water Security and ICID seminar, *What Colour Is Your Water? A critical review of blue, green and other 'waters'* February 2013. University of East Anglia, Water Security Research Centre.

Daigger, G. T. 2009. Evolving urban water and residuals management paradigms: Water reclamation and reuse, decentralization, and resource recovery. *Water Environment Research* 81(8): 809–823.

Fader, M., Gerten, D., Thammer, M., Heinke, J., Lotze-Campen, H., Lucht, W., and Cramer, W. 2011. Internal and external green–blue agricultural water footprints of nations, and related water and land savings through trade. *Hydrology and Earth System Sciences* 15: 1641–1660.

Falkenmark, M. 1995. Coping with water scarcity under rapid population growth. Paper presented at Conference of SADC Ministers, Pretoria, South Africa, November 23–24, 1995.

Falkenmark, M. 2008. Peak water: Entering an era of sharpening water shortage. *Stockholm Water Front* 3–4: 10–11.

Falkenmark, M. and Rockström, J. (2006). The new green and blue water paradigm: Breaking new ground for water resources planning and management. *Journal of Water Resources Planning and Management* 132: 129–132.

FAO. 1996. *Food for All.* Water Food Summit, Rome, November 13–17, 1996. Available online: www.fao.org/docrep/x0262e/x0262e00.htm#TopOfPage (accessed July 2013).

FAO. 2013. The hydrological cycle. Available online: www.fao.org/wairdocs/ilri/x5524e/x5524e04.htm#3.1%20the%20hydrological%20cycle (accessed September 2013).

Gaulke, L. S. 2006. On-site wastewater treatment and reuses in Japan. *Proceedings of the ICE–Water Management* 159(2): 103–109.

Geerts, S., and Raes, D. 2009. Deficit irrigation as an on-farm strategy to maximize crop water productivity in dry areas. *Agricultural Water Management* 96(9): 1275–1284.

Hamdy, A. 2008. Going from rain to gain: Blue and green water management practices. In: A. Santini, N. Lamaddalena, G. Severino, and M. Palladino (Eds.). *Irrigation in Mediterranean Agriculture: Challenges and innovation for the next decades.* Bari: CIHEAM, 7–1 4.

Hoff, H., Falkenmark, M., Gerten, D., Gordon, L., Karlberg, L., and Rockström, J. 2010. Greening the global water system. *Journal of Hydrology*, 384: 177–186.

Karlberg, L., Rockström, J., and Falkenmark, M. 2009. Water resource implications of upgrading rainfed agriculture: Focus on green and blue water trade-offs. Chapter 3 in Suhas P. Wani, Johan Rockstorm, and Theib Oweis (Eds.) *Rainfed Agriculture: Unlocking the potential.* Colombo: CABI International.

Kijne, J. W., Barker, R., and Molden, D. (Eds.) 2003. *Water Productivity in Agriculture: Limits and opportunities for improvement.* Comprehensive Assessment of Water Management in Agriculture Series. Wallingford, Oxon: CABI Publishing.

Kuhn, T. 1962. *The Structure of Scientific Revolutions*, 1st ed., Chicago: University of Chicago.

Lindner, B. 2007. The black water loop: Water efficiency and nutrient recovery combined. Doctoral dissertation. Hamburg: Hamburg University of Technology.

Mekonnen, M. M., and Hoekstra, A. Y. 2011. The green, blue and grey water footprint of crops and derived crop products. *Hydrology and Earth System Sciences* 15: 1577–1600.

Otterpohl, R. 2002. Options for alternative types of sewerage and treatment systems directed to improvement of the overall performance. *Water Science & Technology* 45(3): 149–158.

Otterpohl, R., Oldenburg, M., and Zimmermann, J. 1999. Integrated wastewater disposal for rural settlements. *Wasser und Boden* 51(11): 10–13.

Prathapar, S. 2012. *Color of Water: Is it old wine in a new bottle?* Available online: http://wle.cgiar.org/blogs/2012/11/21/color-of-water-is-it-old-wine-in-a-new-bottle/ (accessed April 25, 2013).

Ringersma, J., Batjes, N., and Dent, D. 2003. *Green Water: Definitions and data for assessment.* Report 2003/3. Wageningen: ISRIC World Soil Information.

Rockström, J. 1999. On-farm green water estimates as a tool for increased food production in water scarce regions. *Physical Chemical Earth* (B) 24(4): 375–383.

Rockström, J., Falkenmark, M., Karlberg, L., Hoff, H., Rost, S., and Gerten, D. 2009. Future water availability for global food production: The potential of green water for increasing resilience to global change. *Water Resources Research* 45, W00A12: 1–16.

Rockström, J., Folke, C., Gordon, L., Hatibu, N., Jewitt, G., Penning de Vries, F., Rwehumbiza, F., Sally, H., Savenije, H., and Schulze, R. 2004. A watershed approach to upgrade rainfed agriculture in water scarce regions through Water System Innovations: An integrated research initiative on water for food and rural livelihoods in balance with ecosystem functions. *Physics and Chemistry of the Earth* 29: 1109–1118.

Rost, S., Gerten, D., Bondeau, A., Lucht, W., Rohwer, J., and Schaphoff, S. 2008. Agricultural green and blue water consumption and its influence on the global water system. Water Resources Research 44: W09405: 1–17.

Savenije, H. 1998. The role of green water in food production in Sub-Saharan Africa. Rome: FAO.

USGS. 2013. The water cycle: Infiltration. Available online: http://ga.water.usgs. gov/edu/watercycleinfiltration.html (accessed September, 2013).

Vidal, A., van Koppen, B., and Blake, D. 2009. *The Green-to-Blue Water Continuum: An approach to improve agricultural systems' resilience to water scarcity.* Waterfront: Stockholm International Water Institute (SIWI).

World Bank 2005. *Shaping the Future of Water for Agriculture: A sourcebook for investment in agricultural water management.* The International Bank for Reconstruction and Development/The World Bank, 1818 H Street NW Washington, DC 20433.

Xu, J. 2013. Effects of climate and land-use change on green-water variations in the Middle Yellow River, China. *Hydrological Sciences Journal* 58(1): 106–117.

Zaag, P. van der, Seyam, I. M., and Savenije, H. H. G. 2002. Towards measurable criteria for the equitable sharing of international water resources. *Water Policy,* 4: 19–32.

# 8 Conclusions

*Jonathan Lautze and Vladimir Smakhtin*

## 8.1 Reviewing Findings

The book is believed to be the first systematic attempt to examine some of the meanings, divergences in interpretation, and value added of water management concepts that have been introduced or grown greatly in use in recent times. Overall, it appears that while the new concepts have brought some benefits, the extent of those benefits may be inflated.

A number of common threads, as well as some differences, can be discerned through a comparison of terms examined in this Book. A summary of possible underlying drivers, value added and sources of confusion for each of the terms considered in previous chapters is presented in Table 8.1. As definitive attribution of underlying factors driving a particular term's creation is somewhat subjective, the word "possible" was added to acknowledge the fact that judgment was involved.

Reviewing underlying drivers across the terms (Table 8.1, column 2) suggests that key language contained in terms was often first used in broader environmental and development discourse and subsequently assimilated into water sector use. This observation indicates that key concepts in water management may often *trickle down* to the water sector rather than *trickling up* in response to real issues specific to the water sector, which in turn spurs questions about the degree to which the water sector leads vs. follows in the selection of key sectoral concepts. There indeed may be timidity when it comes to promulgating key concepts, reflected in a perceived need to couch useful water sector concepts such as water stress or variability management in language such as scarcity and security, respectively.

Moving on to the value obtained through introduction of the new terms (Table 8.1, column 3), at least two general observations are apparent. A first observation is that the value of new terms is more often soft than hard. In other words, soft benefits of awareness-raising, profile-raising, and issue-highlighting would appear to exceed hard benefits such as technical utility. A second observation is that when new terms do possess technical value as an indicator, that value is often exaggerated. There indeed appears a dynamic whereby technical value of an indicator is often trumped by rhetorical value

Table 8.1 New terms—underlying drivers, value added, and sources of confusion

| Term(s) | Possible underlying driver(s) | Value added | Sources of confusion |
|---|---|---|---|
| Water scarcity | Broader focus on natural resources scarcity; desire to identify existing and envisioned regions of "water crisis" | Provides indication of water available to human demand, possibly best utilized in the domestic sector | Multiple prominent definitions of water scarcity; at least three prominent indicators; conflation of water demand and water use when depicting one of the two variables of water scarcity |
| Water governance | Broader focus by development community on governance and good governance | Highlights the role of decision-making processes and institutions in water resources management | Murkiness surrounding the distinction between governance and management; the presumption that good governance necessarily produces good outcomes |
| Water security | Broader use of security language | May provide important package for priority water management concepts | Multiple definitions that are not entirely consistent; paucity of tools that quantify and measure the water security concept |
| Water productivity | Limitations on use of irrigation efficiency at basin scale | When employed alongside other indicators, WP can help guide water allocation decisions | Use of WP in isolation can be misleading; elevating WP to the status of a framework or paradigm rather than an indicator that is best utilized with others |
| Virtual water and water footprint | Desire to understand "hidden" flows of water in food and other commodities; broader environmental footprinting movement | Increases awareness and attention to water used in production, and embedded in traded goods | The concepts are applied as analytical tools and used to generate policy guidance, contexts beyond which they are suited |
| Green and blue water | Desire to simplify esoteric hydrologic language and promote rainfed agriculture | Possible benefits related to raising awareness about role of rainfed agriculture in food security and development | Multiple interpretations of green water; treatment of green water as equivalent to rainfed agriculture |

as a promotional tool, leading to inflated expectations of the value of the indicator.

Reviewing the sources of confusion (Table 8.1, column 4) associated with new term introduction sheds light on two issues. First, it would seem that confusion arises—at least in part—because concepts are deemed to be important despite a lack of clarity on what they mean. The mismatch between presumed importance and unclear meaning would appear to trigger a process of re-positioning traditional concepts and developing and debating new indicators to achieve greater clarity on what terms mean. Second, it would seem that confusion arises due to over-inflation of the capacities of new terms. As already noted, it may be that liberal use of expectation-raising language such as "paradigm," "framework," or "central challenge" has caused terms to be employed in roles beyond which they are suited.

Ultimately, while some skeptics may call for a discussion on whether the aggregate confusion associated with new terms outweighs their collective value added, it may be more constructive to orient discussion toward outlining an improved path forward. In particular, as the flow of new terms is unlikely to stop any time soon, it may be worthwhile to consider: i) how the role, utility, and value of new terms can be kept in context, ii) whether a quality control mechanism can somehow be applied to terms entering water management discourse, iii) whether the process for selecting concepts that are promulgated as central paradigms can be made more explicit.

## 8.2  Moving Forward

### Suggestions for New Term Introduction

Reviewing lessons from past creation of new terms provides a basis for improving the way that new terms are generated in the future. Five recommendations are hereby proposed to strengthen the process of introducing and utilizing new terms.

1.  *Take note of terms that describe terms*   Scrutiny of this book's chapters, as well as many other documents, suggest that key terms themselves have been described with various terminology including: term, concept, indicator, framework, paradigm. While the particular term used to describe a new term has rarely received focus, selection of one appellation vs. another carries key implications for the expectations attached to a term's capacity. It may be inappropriate to attach the label of paradigm or framework, for example, to a simple indicator. A first lesson is therefore to distinguish between term, concept, and indicator on the one hand, and framework or paradigm on the other. Achieving status of framework or paradigm in water management should imply something substantially greater than any of the initial three terms (term, concept, indicator).

2.  *Test whether new terms can be expressed with traditional ones*   Before
    bringing a new term to the attention of the water management
    community, it makes sense to justify the new term's value in relation
    to accepted, traditional terms. If one cannot clearly show that a new
    term provides value over and above existing ones, then there is no need
    to coin that new term. If a term is shown to provide value over and
    above existing terms, the term's creator should nonetheless clearly
    situate the new term relative to traditional concepts so as to minimize
    potential for confusion and conflation between new and old concepts.
3.  *Explicitly state what a particular term excludes and does not do*   Part of the
    confusion and inflated expectations of new terms may be driven by the
    somewhat broad and qualitative definitions often provided for them.
    Such definitions typically highlight what a term can do, but rarely
    highlight what it cannot do. Definitions that fail to reveal what is
    excluded by a concept include those of "IWRM" and "water governance"
    (see Chapter 3). A suggestion is therefore to clearly delimit a term's
    potential uses and role by pointing to issues that a term excludes, and
    roles in which it is not suited. Such an exercise holds substantial
    potential to pre-empt much of the unrealistic expectations and misuse
    of new terms entering water resources management discourse.
4.  *Clearly identify a new term's potential to mislead and be misused*   It would
    seem that many new terms mislead and are misused. One way to limit
    potential for these unwanted "outcomes" is to clearly list areas in which
    terms can mislead, and roles in which terms would be misused.
5.  *Consider creating a process for quality control on new term introduction*   To
    combat the arbitrariness associated with the introduction of new terms
    currently evidenced, it might be beneficial to create a process (e.g., a
    set of criteria, or checklist) for determining that a proposed new term's
    entry into water management discourse generates clear value. While this
    may be done to some extent now through debates contained in the peer
    review process through which terms are proposed and channeled, this
    process may be able to be circumvented by publication in non-peer-
    reviewed outlets. Further, convergence toward clear criteria—for
    example, drawing on those points presented in the lessons immediately
    above—have not been formalized. Formalization of such points could
    clarify the rules of the game, strengthening the peer-review process and
    other scrutiny applied to new term introduction.

### Suggestions for Elevating a Term to Paradigmatic Status

Building on this suggestion to tighten the process of introducing new terms
into water management discourse, it is proposed to put more thought into
selection of central water management paradigms from among terms in water
management. Three criteria are proposed to guide decision-making for
elevating a water management concept to "paradigmatic" status:

- relevance of the concept to water management, as determined for example through aggregation of regional surveys of people active in the water sector;
- potential for the challenges associated with the concept to be solved through options or measures that are water related;
- perception of a concept's potential to serve as a tool that helps mobilize resources.

It is worth clarifying that while a concept's potential to attract resources should come last, it is proposed to make this criterion explicit. As this is clearly a de facto criterion that influences decisions related to which terms grow in prominence, it may be better to acknowledge and formalize it. Nonetheless, it is proposed to *first* outline key concepts in water management based on water sector challenges and the ability to tackle those challenges. With a list of key concepts so established (and presumably revised at some frequency), those determined to be strategic—or more marketable—can ascend to greater prominence.

Five central water management concepts that could result from application of the three criteria stated above are proposed here:

1. *Improving water management*   As noted in Chapter 3, water (resources) management has been defined as "the application of structural and nonstructural measures to control natural and man-made water resources systems for beneficial human and environmental purposes" (Grigg, 1996) and "the study, planning, monitoring, and application of quantitative and qualitative control and development techniques for long-term, multiple use of the diverse forms of water resources" (WHO, 2009, n.p.). While there may be reasons to discard or de-emphasize concepts after a certain number of years of use, certain concepts may remain valid, applicable, and likely more appropriate than new ones that compete to replace them. Valid reasons for replacing a concept may be a concept's anachronism, identification of flaws in a concept, and/or a concept being exceeded in quality by another concept. None of these reasons can be clearly discerned with "water management." The original term could indeed likely work quite well as an encompassing framework or paradigm.

2. *Managing variability (in water availability)*   A fundamental challenge associated with water is variability in its availability, which primarily results from natural variability in rainfall. It would indeed appear that the greater degree of variability in a region or basin, the greater degree of management required to alleviate the negative aspects of that variability. And, conversely, the less variability evidenced, the less management is required. Noticeably, there are obvious in-house (water sector) solutions that can be used to mitigate the effects of variability such as storage augmentation, irrigation expansion, and institutional strengthening. Further, variability may have substantial potential to

resonate popularly so long as concrete examples of droughts, floods, and other extreme weather events are cited. Alleviating negative aspects of water resources variability (e.g., managing extreme droughts and catastrophic floods) while maintaining its benefits (e.g., for satisfying environmental flow requirements) is and will be a primary challenge in the water sector globally.

3.  *Water management for food security*   Agriculture is the largest sectoral use of water. Within agriculture, food production is likely most critical. While this singles out one broader goal of water management, this is likely the single most important outcome of water management. Managing water to enable food security appears a highly relevant goal, and there appear ample water sector options for improvement. The topic is unquestionably marketable. The prominence of the post-2008 hikes in food prices underscores this marketability.

4.  *Management that enables societies to achieve their broader objectives*   Effective water resources management is a means that enables a number of broader ends. Major broader objectives to which water management contributes can be said to include: economic growth, poverty alleviation, agricultural production, environmental sustainability, energy production, among others. Acknowledging linkages between water management and these broader goals is relevant, and presents clear water management response options particularly if water management is viewed as an enabler rather than determinant of broader outcomes.

5.  *Managing conflicts and strengthening institutions*   The fact that water use is a means to many broader objectives—noted immediately above— often creates conflicts between using water for one objective vs. another. In addition, there are often conflicts between use of water in one location vs. another; for example, between an upstream and downstream irrigation scheme, or between an upstream country and a downstream country. Effective institutions constitute the best way to manage conflicts. Rules and procedures, created based on participative processes, that contain equitable and sustainable arrangements for sharing water's benefits across uses and users will help reduce conflicts. This issue certainly is relevant, and there are in-house options such as strengthening water institutions. Further, issues related to conflict appear to attract attention and resources.

## 8.3 Final Thoughts

In conclusion, three messages related to the terminology of water are offered for the water world based on evidence presented in this chapter and this book:

1.  *Clarity in terms' meanings helps reach meaningful decisions that lay the basis for real progress toward real solutions*   Given the degree of challenges

facing the water world, it would seem in our interest to optimally channel efforts so that real issues are tackled. Greater clarity in terms' meanings will reduce time expended sorting through variations in interpretation, and foster greater focus in addressing real challenges.

2. *Constructive discourse to reduce ambiguity surrounding water sector concepts is not a waste of time*   While all terms carry some ambiguity, there is a reason for the existence of text books elaborating fundamental concepts in various disciplines. Common understanding of terms enables us to match words to meanings to enable multiple people to work toward achieving the same commonly understood objective. Conversely, altern-ative understandings of the same term may lead people to pursue alternate objectives contained under the same heading.

3. *There is a need to converge toward agreement on an accepted vocabulary of water sector terms*   While absolute consensus on the meaning and utility of many terms may simply be unachievable, there is nonetheless substantial progress that can be made in converging toward consensus. Various methods could be employed—e.g., participatory processes or surveys of senior people active in the water sector—through which convergence toward common definitions, and meanings of selected terms could be fostered.

## References

Grigg, N. 1996. *Water Resources Management: Principles, regulations and cases*. New York: McGraw Hill.

World Health Organisation (WHO). 2009. Located on Waterwiki.net. Available online: http://waterwiki.net/index.php/Water_management (accessed March 2009).

# Appendix
## Other New Terms in Water Management

*Munir A. Hanjra and Jonathan Lautze*

## 1   New Terms

Many new terms in water management have grown greatly in use in the last two decades. While preceding chapters of this book devoted primary focus to one set of key terms (water scarcity, water governance, water security, water productivity, virtual water, water footprint, green water, blue water) as well as secondary focus to another set (e.g., water stress, economic water scarcity, IWRM), there are no doubt many other terms that merit attention. To give coverage to some of the numerous other new terms that can be found in contemporary water management discussions, this Appendix focuses on an additional set of 25 terms.

To identify new terms, a search was undertaken of 26,952 total sources. The sources included were obtained from databases found in ISI Web of Knowledge (WoK), Scopus, ScienceDirect, EBSCO, Springer, John Wiley, IWA, ASCE, and other water journals; document repositories including IWMI Library Catalogue, organization websites including the World Bank, International Fund for Agricultural Development (IFAD), Consultative Group on International Agricultural Research (CGIAR), International Water Management Institute (IWMI), Asian Development Bank (ADB), African Development Bank (AfDB). A series of searches were performed with the words "water" and "irrigation" selected as search terms in the item title. Bibliographies of identified sources were also searched for additional references. Approximately 260 potential new terms were identified in total.

To filter new terms for inclusion in this Appendix, four criteria were applied. First, terms needed to have a clear relationship to water management. In practice, this often meant that terms possessed a word in their title with a clear water linkage such as "water," "basin," or "downstream." Judgment was nonetheless applied to certain terms to permit their inclusion. Second, there had to be some level of use beyond just one source. In other words, we had to be clear that the term is in fact put into use by people beyond the author that coined a term. Third, there had to be evidence that frequency of a term's use had grown substantially in the last two decades. Last, terms found in previous chapters were removed from the list.

The remainder of this Appendix focuses simply on provision of a short definition and explanation of the meaning of each new term. Due to limitations on the space in which each term is defined and discussed, reference to two notable works related to each term are provided immediately after initial mention of each term. This will enable readers to follow up on their own should they wish to gain a more in-depth understanding of a particular term.

It should be noted that the style of this Appendix is an admitted departure from the book's main chapters. The sheer number of terms receiving focus in this Appendix motivated a need to adapt the approach utilized in Chapters 2 through 7. That is, whereas Chapters 2 through 7 *analyze* one or very few terms, this Appendix *describes* many terms. As such, the Appendix provides a descriptive review to give general coverage to the set of terms rather than an analysis of their value added. A short final section nonetheless offers some thoughts about the collective benefits achieved through development of the reviewed terms.

1. *Downstreamness* (van Oel et al., 2011; van Oel and van der Veen, 2011) In many river basins, rising demands for freshwater have led to the development of water storage and irrigation infrastructure, which raise tensions and trade-offs between upstream and downstream users and affect the choice of technical interventions for water resources development and management. The term downstreamness was proposed for analyzing the availability and commitments of freshwater in river basins from the perspective of a riparian's position (van Oel et al., 2011). Downstreamness of a location is the ratio of its upstream catchment area to the entire river basin area. The downstreamness of a water-related function (e.g., hydropower) is defined as the downstreamness-weighted integral of that function divided by its regular integral. The concept has been used to determine the downstreamness of surface water storage capacity, stored surface water volumes, and water demand. In the context of upstream vs. downstream groundwater management, a downstreamness perspective can help to reveal whether interventions such as rainwater harvesting lead to uneven recharge of groundwater (Ray and Bijarnia, 2006).

2. *Ecological sanitation* (Langergraber and Muellegger, 2005; Haq and Cambridge, 2012) The term ecological sanitation (or Ecosan, as it is often abbreviated) refers to crop fertilization with human urine to increase water and nutrient availability in rainfed smallholder agriculture (Andersson et al., 2011). It embodies the potential impact of two strategies—namely, i) in situ water harvesting and ii) fertilization with human urine—to increase agricultural production and promote more environmentally friendly sanitation. The recycling of nutrients embedded in human excreta for use in agriculture is therefore seen as

a potential strategy to both enhance soil fertility in smallholder systems and address sanitation challenges. Nevertheless, the benefits are not always straightforward and the implementation of ecological sanitation faces diverse issues such as changes in crop yield, health risks, economic viability, and social perceptions and adoption challenges (Andersson et al., 2011).

3. *Green economy* (UN-Water, 2012; Kadekodi, 2013)   Green economy refers to a world in which natural resources including forests, water, land, and ecosystems are sustainably managed and conserved to improve livelihoods and ensure food security. Healthy ecosystems created by such an approach are in turn presumed to increase the economic returns from the activities they support (World Bank, 2013). The term green economy is consistent with objectives of lowering the carbon footprint and greenhouse gas emissions and making the development process more environmentally friendly and socially responsible. A green economy is presumed to depend on sustainable, integrated, and more efficient management of water resources; safe and sustainable provisioning of water supply and sanitation services; and effective management of water variability, ecosystems, and the resulting impacts on livelihoods (UN-Water, 2012). Building a robust green economy is considered critical in the context of a changing climate.

4. *Greening the global water system* (Hoff, 2009; Hoff et al., 2010)   While "greening" can be utilized to refer to a range of environmentally friendly water resources management measures (e.g., Villanueva et al., 2003), the notion of greening the global water system can be more specifically tied to recognition for the role of green water in food production and for seeking solutions to water management challenges through green water options. Thus it advocates for more emphasis on green water in the context of an integrated approach to the green to blue water spectrum.[1] Greening the global system extends interventions beyond more conventional approaches of increasing irrigated area to place focus on rainwater harvesting, supplemental irrigation, vapor shift, and soil and nutrient management to upgrade rainfed systems (Rockström, 2003). Such strategies are said to open up new avenues for water management for sustainable development and poverty alleviation (Hoff et al., 2010).

5. *Hydrocentricity* (Brichieri-Colombi, 2004; Allan, 2007)   The term hydrocentricity is used to characterize business-as-usual approaches to meeting society's needs for goods and services that rely on conventional "in-the-box" water management frameworks and interventions (Brichieri-Colombi, 2004). Hydrocentricity is the narrow lens through which water resource planners have traditionally resolved water

management problems. Proponents of hydrocentricity language typically advocate for a broader perspective, explaining that a wider set of solutions to such water problems can be identified through non hydrocentric lenses. Thus, instead of creating an artificial boundary at the river basin, there is a need to expand the planning space beyond the water sphere to resolve water management issues.

6. *Hydrocracy* (Molle et al., 2009; Wester et al., 2009) A hydrocracy can be simply defined as a water bureaucracy (Molle et al., 2009). Hydrocracies can be described as bureaucratic organizational frameworks that not only operate as neutral control bodies, but that are also driven by their own interests. Hydrocracies consist of technically and economically oriented engineers, technical employees of state or federal departments, ministries, and state secretaries concerned with water issues. Members of hydrocracies form an important part of epistemic communities, setting water agendas through use of discourses and political influence (Treffner et al., 2010).

7. *Hydro-hegemony* (Jagerskog, 2008, and associated special issue of *Water Policy*) In the context of international relations, hegemony is understood as authoritative leadership imposed by one powerful actor, such as a state, over a weaker one. In water management, hydro-hegemony usually refers to hegemonic interaction over transboundary water resources in river basins shared by two or more nations; however, it can be utilized for the exertion of power and control within a state (Zeitoun and Warner, 2006; Jagerskog, 2008). Control by the basin hegemony may be exerted in the form of resource capture or containment of challenges (through exploitation of infrastructure development and potential) of riparian position, and of power asymmetries. Manifestations of hydro-hegemony range from coercive cooperation with an inequitable distribution of benefits derived from resource use, to domination (Treffner et al., 2010).

8. *Hydropolitics* (Salman and Uprety, 1999; Movik, 2010) The term hydropolitics refers to the political reality that geographic boundaries in most nation states are politically determined and these boundaries generally do not align with apolitical hydrologic boundaries. Countries' independence therefore creates difficult issues for sharing the water resources. Examples include Pakistan's independence in 1947, which led to the protocols and agreements that were incrementally and painstakingly negotiated for establishing the Indus Basin Commission under the Indus Basin Treaty for the sharing of water resources among the Indus basin states, principally India and Pakistan (Wescoat Jr. et al., 2000). Similarly, with the independence of the Central Asian

states from the Soviet Union, hydraulic infrastructure began to have transboundary implications such that cooperative institutional mechanisms for the operation and management of these waterworks needed to be negotiated.

9. *Hydrosolidarity* (Pigram, 2000; Kemerink et al., 2009) The term hydrosolidarity refers to the commonality of interests and mutual dependence required to achieve and maintain the health and vitality of a river basin as a trade-off against economic viability and regional development (Pigram, 2000). Different reaches across a river basin often represent a mosaic of contrasting and competing biophysical, economic, and social circumstances and community interests, and even irrigation mosaics (Paydar et al., 2011). Thus water and health of the river basin is the common thread and unifying element that promotes communality of attitude toward the river system across reaches. A case example is Australia's Murray Darling Basin, where a degree of hydrosolidarity is emerging due to water over allocation and pressing needs for coordinated resource management across the basin.

10. *Land grabs and water grabs* (Williams et al., 2012; Lagerkvist, 2013) Most countries in the developing world have water rights defined in terms of land rights such that the transfer of land, either through sale or acquisitions through political and customary systems by third parties, results in transfer of water rights. This has implications for large-scale investments in land and associated transfer of land and water rights. Such "transfer" of land and water has often been colloquialized as land and water "grabs." The terms land grabs and water grabs are reflected in recent large-scale land acquisitions for biofuel production by international investors, and foreign governments for assuring food security for fast-growing populations in their land-scarce countries by investing in large-scale land deals offshore (e.g., many Middle East countries, China, and even India investing in land-abundant African countries such as Ethiopia, Ghana, Mozambique, Zimbabwe). Some have contended (Williams et al., 2012) that land and water "grabs" should not be treated as inherently negative, given that such grabs reflect foreign investment that holds potential to deliver a multitude of benefits such as new income, employment, and livelihood opportunities.

11. *Multiple use systems* (Renwick, 2001; Senzanje et al., 2008) The term multiple use systems (MUS) dispels the misperception that irrigation systems are designed for supplying water only to crops, and denotes the fact that almost all irrigation systems simultaneously provide water for other uses (Meinzen-Dick and van der Hoek, 2001). MUS means that water has multiple uses, and irrigation infrastructures are in fact multiple

use systems. Irrigation systems not only supply water for the growing of main crops, but also water for home gardens, trees and agroforestry, permanent vegetation, livestock, poultry etc. Viewing irrigation infra-structure as multiple use systems has implications for water management and water policy. It can enhance the economic and social value of water in irrigation systems (Renwick, 2001), can lead to more productive and environmentally sustainable systems, promote recognition for the rights of all users and promote more equitable and socially just outcomes.

12. *Natural infrastructure* (Emerson and Bos, 2004; IUCN, 2011)   Natural infrastructure (sometimes called green infrastructure) has been defined by the US Environmental Protection Agency (US EPA, 2013) as the interconnected network of natural and undeveloped areas needed to maintain and support ecosystems. Natural infrastructure has also been defined as "another term used to describe ecosystem services—the benefits that people obtain from nature" (IUCN, 2011). Watersheds themselves are natural infrastructure in the sense that they perform infrastructure functions for water. Other examples of natural infrastruc-ture include "upland soils that store water, wetlands that store and clean water, floodplains that buffer floods, rivers that provide water conveyance and mangroves, coral reefs and barrier islands that protect coastal communities" (IUCN, 2011). It appears the underlying driver for creation of the concept was the desire to increase attention for the role of ecosystems in fulfilling functions of built infrastructure. Indeed, with the term infrastructure defined as "the stock of facilities, services and installations needed for the functioning of a society," nature is part of the infrastructure portfolio of every country and every economy. Nature is then "natural infrastructure" based on its capacity to comple-ment, augment, or even replace the services provided by traditional engineered infrastructure (Emerson and Bos, 2004).

13. *Problemshed* (Mollinga et al., 2007)   Problemshed can be viewed as a construct that is strategically delimited in such a way as to best address a problem. It is argued that conventional policy and institutional reforms in agricultural water management are characterized by social engineering approaches that fail to address the inherently political nature of the reform processes, or their embeddedness. Such acknowledgement leads to rethinking of the unit of analysis and refinements in policy reform, which is captured in the notion of problemsheds. Thus, the idea of viewing a river basin as a closed system in anything but hydrological terms as a "watershed" is a reflection of the concept that has rather grown as a "problemshed" (Brichieri-Colombi, 2004).

14. *Resilience* (Young, 2010; Silici et al., 2011)   Resilience definitions can be broadly divided into two groups: i) the capacity to absorb or withstand

disturbances or shocks, which is very similar to use of the word "resistance," ii) the capacity to recover from shocks due to the regenerative abilities of a social- or an eco-system. The former interpretation is focused more on the capability of a system to maintain its basic functions and structures in a time of shocks and perturbations such that resources and ecosystem services essential for human livelihoods continue to be delivered (Adger et al., 2005; Allenby and Fink, 2005). The latter interpretation places more emphasis on the ability to learn and adapt to changes and shocks (Birkmann, 2006). Resilience has been described as the opposite of vulnerability (Adger et al., 2005). In the context of water management, resilience has been associated with storing water to reduce the effects of erratic rainfall (Brown and Lal, 2006).

15. *River basin closure* (Falkenmark and Molden, 2008; Molle et al. 2010) When the supply of water falls short of commitments to fulfill demand in terms of water quality and quantity in a basin, for part or all of the year, that basin is said to be closed (Falkenmark and Molden, 2008; Molle et al., 2010). River basin closure denotes that increasing water withdrawals for urban, industrial, and agricultural uses have left little water for the environment and profoundly altered the hydrology and natural flows of many river basins worldwide. As a result of increasing water withdrawals, a growing number of river basins are closing. Examples include many river basins around the globe such as the Colorado, Yellow River basin (China), Amu Darya, Syr Darya, and the Jordan (Molle, 2008).

16. *River basin trajectories* (Venot et al., 2008; Molle and Wester, 2009) River basin trajectories refer to the historical paths of basin development. The trajectories concept implies that basin development history in most countries is a response to certain generic conditions in the agriculture sector and more broadly in society (Venot et al., 2008; Smedema, 2011). When such conditions are absent, very little development takes place and the development trajectory is near static—even when there is an evident need for such development. These generic conditions include: state of agricultural development; sector outlook and perspectives; societal priorities and need for food security; employment and population settlement (Smedema, 2011).

17. *Sanitation ladder* (Drechsel et al., 2010; Keraita et al., 2010) The term sanitation ladder refers to the spectrum of approaches—from low-tech to high-tech—used for risk reduction through treatment in wastewater irrigated agriculture. Less developed countries typically employ approaches at the lower end of the sanitation ladder while more developed countries employ approaches at the higher end of the sanitation ladder. Higher end options generally include wastewater treatment,

which puts emphasis on the regulation of water quality standards. Lower tech options are less focused on water quality regulation and more focused on reducing risks to human health. Multiple barrier approaches are those in which a combination of measures—e.g., safe farm-based and post-harvest measures, consumer education, and social incentives to farmers—is employed based on their suitability in a given context (Keraita et al., 2010).

18. *Water Accounting* (Molden, 1997; Karimi et al., 2013)   Water Accounting (WA) is a procedure developed by IWMI for documenting the uses, depletion, and productivity of water in a water basin context (Molden, 1997). The United Nations Statistics Division has also proposed a water accounting framework called System of Environmental Economic Accounting for Water (SEEAW). SEEAW describes hydrological and economic information through a set of standard tables; with supplementary tables to cover social aspects (UN, 2007). SEEAW accounting includes precipitation, soil water, and natural evapotranspiration. More recently, Water Accounting Plus (WA+) is a framework designed to provide explicit spatial information on water depletion and net withdrawal processes in complex river basins. In the WA+ framework, the influence of land use and landscape ET on the water cycle is described explicitly by defining land use groups with common characteristics (Karimi et al., 2013). Fundamental differences between WA and WA+ include, but are not limited to, i) greater reliance on satellite-based data in WA+, and ii) refinement of water depletion.

19. *Water banking (institutional)* (Clifford et al., 2004; Lepper, 2006)   Water banking is the practice of forgoing water deliveries during certain periods, and "banking" the right to use the forgone water in the future, or saving it for someone else to use in exchange for a fee (O'Donnell and Colby, 2010). Water banking typically occurs through an institutionalized process with known procedures and some measure of public sanction for its activities (Clifford, 2004). Water banking occurs in dry areas of the world such as the American West. In Arizona, for example, the Arizona Water Banking Authority (AWBA) banks water in upstream Colorado for use during dry periods (AWBA, 2013).

20. *Water banking (groundwater banking)* (Purkey et al., 1998; Pulido-Velazquez et al. 2004)   "Groundwater banking" can be considered one type of water banking. Groundwater banking has come to refer to the practice of storing excess water supplies in subsurface aquifers so that they can later be withdrawn (Purkey et al., 1998). Brothers and Katzer (1990) provide an example of groundwater banking through artificial recharge in Nevada, USA, for example, and Karimov et al. (2010)

provide an example of groundwater banking through managed aquifer recharge in wet periods for use in dry periods in the Fergana Valley of Central Asia.

21. *Water buy-backs* (Zaman et al. 2009; Wheeler et al. 2013)    The term water buy-backs refers to the market-based acquisitions of water from irrigators by the government or a third party acting on behalf of the environment and releasing this water back to the environment to improve environmental flows and health of the river system (Australian Government, 2012). The concept of water buy-backs implies that water was taken from the environment for agricultural uses and this led to over-abstractions with consequences for environmental assets and ecosystem functions. As such, water now should be bought back and released back to the environment. A case example is found within the Water for the Future Program of the Australian Government. Where environmental requirements are not met, the Australian Government can buy additional water back from the irrigators through direct participation in the water market, but the sales are purely voluntary and water is held by the national regulator for use by the environment (MDBA, 2012).

22. *Water diplomacy* (Kirmani and Le Moigne, 1997; Dinar and Dinar, 2000; Islam and Susskind, 2012)    Water diplomacy has been considered as all contact concerning transboundary waters that involves at least one state actor (van Genderen and Rood, 2011). Water diplomacy has been called "a theory and practice of implementing adaptive water management for complex water issues" (Islam and Susskind, 2012; Tufts University, 2013). Dinar and Dinar (2000) discuss water diplomacy in the context of a few transboundary river basins, through which one might deduce their interpretation of water diplomacy is the process of negotiations surrounding transboundary waters. Ultimately, one interprets that definitions and use of "water diplomacy" are quite broad.

23. *Water poverty and water poverty index* (Sullivan et al., 2003; Forouzani et al., 2013)    Lawrence et al. (2002) stated that "people can be said to be 'water poor' in the sense of not having sufficient water for their basic needs because it is not available." Water poverty would appear an extrapolation of this. An index of water poverty was created that consisted of five components: i) resources, ii) access, iii) capacity, iv) use, v) environment (Sullivan et al., 2003). Each of these components contains several subcomponents. The Water Poverty Index is based on the architecture of the Human Development Index, and many recent examples of its application exist (Komnenic et al., 2009; Cho et al., 2010; Forouzani et al., 2013).

24. *Water towers* (Immerzeel et al., 2008; UNEP, 2013)    The term "water tower" originally referred to "a tower supporting an elevated tank, whose height creates the pressure required to distribute the water through a piped system" (Soanes and Stevenson, 2004). However, in the context of hydrology and river basin management, it is used as a symbolic term for a mountain area that supplies disproportionate runoff as compared to surrounding lowland areas (Viviroli et al., 2007). The concept is therefore a relative one. In Africa, water towers have been identified as areas such as the Fouta Djallon in Guinea, the Blue Nile basin in Ethiopia, and the Angolan Plateau in northern Angola (UNEP, 2013). These are areas where relatively large quantities of water can be said to originate.

25. *Water–Food–Energy nexus* (McCornick et al., 2008; Stillwell et al. 2011) A nexus is a connection or set of connections between two or more things. A nexus approach is considered an approach that integrates management and governance across sectors and scales. A nexus approach between water, food, and energy seeks to recognize that water is required for both food and energy, and there are benefits to acknowledging this and considering both from a water perspective. It also acknowledges that water–food–energy insecurities may be best tackled through a cross-sectoral approach to enhancing infrastructure and access. It would indeed appear that opportunities to raise productivity and efficiency— and to reduce vulnerability to rainfall variability and climate change —are optimally implemented through approaches that simultaneously consider water requirements for the multiple sectors for which the resource plays a central role (McCornick et al., 2008; Hoff, 2011). This entails consideration for major trade-offs among the food, energy, and environmental goals of water and energy development, allocation, and management.

## 2    Closing Thoughts

Reviewing the terms outlined above, some have likely provided more value than others. While the descriptive nature of this chapter constrains definitive assignment of the aggregate value achieved through development of these terms, it seems safe to say that many of the new terms presented have in fact generated some value. Several of the terms—natural infrastructure, the water–food–energy nexus, land and water grabs—have added an allure to issues that may have otherwise been neglected. Other terms—e.g., water accounting, water poverty—offer new ways to measure water use and access. Still other terms—e.g., hydro-hegemony, river basin closure, river basin trajectory—provide conceptual lenses that may help clarify dynamics surrounding basin-level water use.

Despite the value achieved through use of certain terms, it is likely that not all terms generated major new value. Hydropolitics and hydrodiplomacy, for example, could likely be replaced by water politics and water diplomacy, respectively. Further, even in cases where there is value in a new term, one needs to pose questions concerning whether that term's value outweighs its possible negative attributes such as the increased confusion. The analytical depth to which terms were scrutinized in this Appendix is simply insufficient to address this issue. Nonetheless, future independent work could be undertaken to rigorously compare individual terms presented in this Appendix against pre-existing terms, in order to identify their value added.

Assessment of the underlying drivers for creating these terms is similarly constrained by the review-nature of this chapter. Nonetheless, if one were to conjecture a bit, it would seem that creation of many of these terms was driven by the desire to package pre-existing concepts under a trendier heading of fewer words. "Water bureaucracy" was merged to one, sexier word in "hydrocracy." Saving water during one season for use in another was condensed into the label of "water banking." Areas of relative water abundance, typically in mountainous areas, were given the catchy appellation of "water towers." While the rationale of condensing a long set of terms into a short set designed to attract interest is not unreasonable, it may at the same time be important to keep in context the value that these terms add.

Overall, this Appendix has attempted to capture and give coverage to some of the numerous terms in water management that did not receive attention in the book's preceding chapters. It is hoped that this Appendix will help catalyze interest in more in-depth future investigation of the value that these new terms have added. Despite the analytical constraints on this Appendix, largely a byproduct of the number of terms explored, the Appendix may nonetheless offer suggestive evidence that is consistent with the preceding chapters. That is, it appears that while many of the new terms presented have provided value, the overall amount of value that they have generated must be kept in context. It indeed appears that many terms are simplifications of issues already known and traditionally described with a greater number of words.

## Note

1. See Chapter 7 of this book for a more detailed explanation of green and blue water.

## References

Adger, W. N., Hughes, T. P., Folke, C., Carpenter, S. R., and Rockström, J. 2005. Social–ecological resilience to coastal disasters. *Science* 309: 1036–1039.

Allan, J. A. 2007. Beyond the watershed: Avoiding the dangers of hydrocentricity and informing water policy. *Water Resources in the Middle East* 2: 33–39.

Allenby, B., and Fink, J. 2005. Towards inherently secure and resilient societies. *Science* 309: 1034–1036.

Andersson, J. C. M., Zehnder, A. J. B., Rockström, J., and Yang, H. 2011. Potential impacts of water harvesting and ecological sanitation on crop yield, evaporation and river flow regimes in the Thukela River basin, South Africa. *Agricultural Water Management* 98(7): 1113–1124.

Arizona Water Banking Authority (AWBA). 2013. Arizona Water Banking Authority. Available online: www.azwaterbank.gov

Australian Government. 2012. *Restoring the Balance in the Murray-Darling Basin: Water entitlement purchasing in the Murray-Darling Basin.* Canberra: Department of Environment, Australian Government.

Birkmann, J. 2006. Measuring vulnerability to promote disaster-resilient societies: Conceptual framework and definitions. Chapter 1 in: J. Birkmann (Ed.) *Measuring Vulnerability to Natural Hazards: Towards disaster resilient societies.* Tokyo: United Nations University Press, 9–54.

Brichieri-Colombi, J. S. 2004. Hydrocentricity: A limited approach to achieving food and water security, *Water International* 29(3): 318–328.

Brothers, K., and Katzer, T. 1990. Water banking through artificial recharge. Las Vegas Valley, Clark County, Nevada. *Journal of Hydrology* 115: 77–103.

Brown, C., and Lal, U. 2006. Water and economic development: The role of variability and a framework for resilience. *Natural Resources Forum* 30: 306–317.

Cho, D., Ogwang, T., and Opio, C. 2010. Simplifying the water poverty index. Social Indicators Research 97: 257–267.

Clifford, P., Landry, C., and Larsen-Hayden, A. 2004. *Analysis of Water Banks in the Western States.* Washington, DC: Washington Department of Ecology, Publication number 04–11–011.

Dinar, S., and Dinar, A. 2000. Negotiating in international watercourses: Diplomacy, conflict, and cooperation. *International Negotiation* 5: 193–200.

Drechsel, P., Scott, C. A., Raschid-Sally, L., Redwood, M., Bahri, A. (Eds.). 2010. *Wastewater Irrigation and Health: Assessing and mitigating risk in low-income countries.* Colombo, Sri Lanka: IWMI; London: Earthscan; Ottawa: International Development Research Centre (IDRC).

Emerton, L., and Bos, E. 2004. *Value: Counting ecosystems as an economic part of water infrastructure.* Gland, Switzerland and Cambridge, UK: IUCN.

Falkenmark, M., and Molden, D. (2008) Wake up to realities of river basin closure. *International Journal of Water Resources Development* 24(2): 201–215.

Forouzani, M., Karami, E., and Zamani, G. H. 2013. Agricultural water poverty in Marvdasht County, Southern Iran. *Water Policy* 15(5): 669–690.

Haq, G., and Cambridge, H. 2012. Exploiting the co-benefits of ecological sanitation: Current opinion in environmental sustainability 4(4): 431–435.

Hoff, H. 2009. Global water resources and their management. *Current Opinion on Environmental Sustainability* 1(2): 141–147.

Hoff, H. 2011. *Understanding the Nexus.* Background Paper for the Bonn 2011 Conference: The Water, Energy and Food Security Nexus. Stockholm: Stockholm Environment Institute.

Hoff, H., Falkenmark, M., Gerten, D., Gordon, L., Karlberg, L., and Rockström, J. 2010. Greening the global water system. *Journal of Hydrology: Green–Blue Water Initiative (GBI)* 384(3–4): 177–186.

Immerzeel, W., Stoorvogel, J., and Antle, J. 2008. Can payments for ecosystem services secure the water tower of Tibet? *Agricultural Systems* 96 (1–3): 52–63.

Islam, S., and Susskind, L. 2012. *Water Diplomacy: A negotiated approach to managing complex water networks.* New York and London: RFF Press.

IUCN. 2011. *Natural Infrastructure: Part of the hydropower debate.* Available online: www.iucn.org/news_homepage/news_by_date/?7744/Natural-infrastructure-part-of-the-hydropower-debate

Jakerskog, A. 2008. Prologue: A special issue on hydro-hegemony. *Water Policy* 10 Supplement 2: 1–2.

Kadekodi, G. 2013. Is a "Green Economy" possible? *Economic and Political Weekly* 48 (25).

Karimi, P., Bastiaanssen, W. G. M., and Molden, D. 2013. Water Accounting Plus (WA+): A water accounting procedure for complex river basins based on satellite measurements. *Hydrology and Earth Systems Sciences* 17: 2459–2472.

Karimov, A., Smakhtin, V., Mavlonov, A., and Gracheva, I. 2010. Water "banking" in Fergana valley aquifers: A solution to water allocation in the Syrdarya river basin? *Agricultural Water Management* 97: 1461–1468.

Kemerink, J. S., Ahlers, R., van der Zaag, P. 2009. Assessment of the potential for hydro-solidarity within plural legal conditions of traditional irrigation systems in northern Tanzania. *Physics and Chemistry of the Earth* Parts A/B/C 34 (13–16): 881–889.

Keraita, B., Drechsel, P., and Konradsen, F. 2010. Up and down the sanitation ladder: Harmonizing the treatment and multiple-barrier perspectives on risk reduction in wastewater irrigated agriculture. *Irrigation and Drainage Systems* 24(1–2): 23–35.

Kirmani, S., and Le Moigne, G. 1997. *Fostering Riparian Cooperation in International River Basins: The World Bank at its best in development diplomacy.* The World Bank. Available online: file://C:%5CCcgiar%5CCiwmi%5COldMatter2004%5CElizabeth 2006b@Lit%5CFostering%20Riparian%20Cooperation%20in%20International %20River%20Basins.pdf

Komnenic, V., Ahlers, R., and Zaag, P. V. D. 2009. Assessing the usefulness of the water poverty index by applying it to a special case: Can one be water poor with high levels of access? *Physics and Chemistry of the Earth* Parts A/B/C 34(4–5): 219–224.

Lagerkvist, J. 2013. As China returns: Perceptions of land grabbing and spatial power relations in Mozambique. *Journal of Asian and African Studies.* doi:10.1177/ 0021909613485217

Langergraber, G., and Muellegger, E. 2005. Ecological sanitation: A way to solve global sanitation problems? *Environment International* 31(3): 433–444.

Lawrence, P., Meigh, J., and Sullivan, C. 2002. *The Water Poverty Index: An international comparison.* Keele Economics Research Papers. Keele University.

Lepper, T. 2006. Banking on a better day: Water banking in the Arkansas Valley. *The Social Science Journal* 43(3): 365–374.

McCornick, P. G., Awulachew, S. B., and Abebe, M. 2008. Water–food–energy– environment synergies and tradeoffs: Major issues and case studies. *Water Policy* 10: 23–30.

Meinzen-Dick, R., and van der Hoek, W. 2001. Multiple uses of water in irrigated areas. *Irrigation and Drainage Systems* 15(2): 93–98.

Molden, D. 1997. *Accounting for Water Use and Productivity*, SWIM Paper 1. Colombo, Sri Lanka: IWMI.

Molle, F. 2008. Why enough is never enough: The societal determinants of river basin closure. *International Journal of Water Resources Development* 24(2): 217–226.

Molle, F., and Wester, P. 2009. *River Basin Trajectories: Societies, environments and development*. Wallingford, Oxon: CABI Publishing.

Molle, F., Mollinga, P. P., and Wester, P. 2009. Hydraulic bureaucracies and the hydraulic mission: Flows of water, flows of power. *Water Alternatives* 2(3): 328–349.

Molle, F., Wester, P., and Hirsch, P. 2010. River basin closure: Processes, implications and responses. *Agricultural Water Management* 97(4): 569–577.

Mollinga, P. P., Meinzen-Dick, R. S., Merrey, D. J., 2007. Politics, plurality and problemsheds: A strategic approach for reform of agricultural water resources management. *Development Policy Review* 25(6): 699–719.

Movik, S. 2010. Return of the Leviathan? "Hydropolitics in the developing world" revisited. *Water Policy* 12(3): 641–653.

Murray Darling Basin Authority (MDBA). 2012. Guide to the Proposed Basin Plan, Vol. 1 and Vol. 2. Canberra: Murray Darling Basin Authority, Australian Government.

O'Donnell, M., and Colby, B. 2010. *Water Banks: A tool for enhancing water supply reliability*. The University of Arizona Department of Agricultural and Resource Economics.

Paydar, Z., Cook, F., Xevi, E., and Bristow, K. 2011. An overview of irrigation mosaics. *Irrigation and Drainage* 60(4): 454–463.

Pigram, J. J. 2000. Towards upstream–downstream hydrosolidarity: Australia's Murray-Darling river basin. *Water International* 25(2): 222–226.

Pulido-Velazquez, M., Jenkins, M. W., and Lund, J. R. 2004. Economics values for conjunctive use and water banking in Southern California. *Water Resources Research* 40(3): W03401.

Purkey, D., Thomas, G., Fullerton, D., Moench, M., and Axelrad, L. 1998. *Feasibility Study of a Maximal Program of Groundwater Banking*. San Francisco, CA: Natural Heritage Institute.

Ray, S., and Bijarnia, M. 2006. Upstream vs downstream: Groundwater management and rainwater harvesting. *Economic and Political Weekly* 41(23): 2375–2383.

Renwick, M. E. 2001. Valuing water in a multiple-use system: Irrigated agriculture and reservoir fisheries. *Irrigation and Drainage Systems* 15(2): 149–171.

Rockström, J. 2003. Water for food and nature in drought-prone tropics: Vapour shift in rain-fed agriculture. *Philosophical Transactions of the Royal Society B: Biological Sciences* 358(1440): 1997–2009.

Salman, S. M. A., and Uprety, K. 1999. Hydropolitics in South Asia: A comparative analysis of the Mahakali and the Ganges Treaties. *Natural Resources Journal* 39(2): 295–343.

Senzanje, A., Boelee, E., and Rusere, S. 2008. Multiple use of water and water productivity of communal small dams in the Limpopo Basin, Zimbabwe. *Irrigation and Drainage Systems* 22(3–4): 225–237.

Silici, L., Ndabe, P., Friedrich, T., and Kassam, A. 2011. Harnessing sustainability, resilience and productivity through conservation agriculture: The case of Likoti in Lesotho. *International Journal of Agricultural Sustainability* 9(1): 137–144.

Smedema, L. K. 2011. Drainage development: Driving forces, conducive conditions and development trajectories. *Irrigation and Drainage* 60(5): 654–659.

Soanes, C., and Stevenson, A. (Eds.). 2004. *The Concise Oxford English Dictionary.* Oxford: Oxford University Press.

Stillwell, A. S., King, C. W., Webber, M. E., Duncan, I. J., and Hardberger, A. 2011. The energy–water nexus in Texas. *Ecology and Society* 16 (1): 2.

Sullivan, C., Meigh, J., and Giacomello, A. 2003. The water poverty index: Development and application at a community scale. *Natural Resources Forum* 27: 189–199.

Treffner, J., Mioc, V., and Wegerich, K. 2010. A–Z Glossary in: K. Wegerich, and J. Warner (Eds.) *Politics of Water: A survey.* London and New York, Routledge, 215–320.

Tufts University. 2013. Water Diplomacy. Available online: https://sites.tufts.edu/water diplomacy/program

UN. 2007. *System of Environmental Economic Accounting for Water.* Geneva: United Nation Statistics Division.

UN-Water. 2012. *Water in a Green Economy.* A statement by UN-Water for the UN Conference on Sustainable Development 2012 (Rio+20 Summit).

UNEP. 2013. *Africa Environment Outlook 3: A Summary for Policy-Makers.* UNEP. Nairobi, Kenya. Available online: www.unep.org/pdf/aeo3.pdf

United States Environmental Protection Agency. 2013. *Natural Infrastructure.* Available online: www.epa.gov/region03/green/infrastructure.html

van Genderen, R., and Rood, J. 2011. *Water Diplomacy: A niche for the Netherlands?* The Netherlands: Ministry of Foreign Affairs.

van Oel, P., Krol, M., and Hoekstra, A. 2011. Downstreamness: A concept to analyze basin closure. *Journal of Water Resources Planning and Management* 137(5): 404–411.

van Oel, P. R., and van der Veen, A. 2011. Using agent-based modeling to depict basin closure in the Naivasha basin, Kenya: A framework of analysis. *Procedia Environmental Sciences* 7(0): 32–37.

Venot, J.-P., Biggs, T., Molle, F., and Turral, H. 2008. Reconfiguration and closure of river basins in south India: Trajectory of the lower Krishna basin. *Water International* 33(4): 436–450.

Villanueva, L. C., Delatorre, J., De la Riva, F., and Monardes, V. 2003. Greening of arid cities by residual water reuse: A multidisciplinary project in Northern Chile. *AMBIO: A Journal of the Human Environment* 32(4): 264–268.

Viviroli, D., Dürr, H. H., Messerli, B., Meybeck, M., and Weingartner, R. 2007. Mountains of the world, water towers for humanity: Typology, mapping, and global significance. *Water Resources Research* 43, W07447.

Wescoat Jr, J. L., Halvorson, S. J., and Mustafa, D. 2000. Water management in the Indus basin of Pakistan: A half-century perspective. *Water Resources Development* 16(3): 391–406.

Wester, P., Rap, E., and Vargas-Velázquez, S. 2009. The hydraulic mission and the Mexican hydrocracy: Regulating and reforming the flows of water and power. *Water Alternatives* 2(3): 395–415.

Wheeler, S., Loch, A., Zuo, A., and Bjornlund, H. 2013. Reviewing the adoption and impact of water markets in the Murray-Darling Basin, Australia. *Journal of Hydrology.* doi:http://dx.doi.org/10.1016/j.jhydrol.2013.09.019

Williams, T. O., Gyampoh, B., Kizito, F., and Namara, R. 2012. Water implications of large-scale land acquisitions in Ghana. *Water Alternatives* 5(2): 243–265.

World Bank. 2013. *Sharing Smart Solutions in Water: 2012 Annual Report and Phase 1 Summary.* Washington, DC: World Bank, Water Partnership Program.

Young, O. R. 2010. Institutional dynamics: Resilience, vulnerability and adaptation in environmental and resource regimes. *Global Environmental Change* 20(3): 378–385.

Zaman, A. M., Malano, H. M., Davidson, B. 2009. An integrated water trading-allocation model, applied to a water market in Australia. *Agricultural Water Management* 96(1): 149–159.

Zeitoun, M., and Warner, J. 2006. Hydro-hegemony: A framework for analysis of trans-boundary water conflicts. *Water Policy* 8: 435–460.

# Index